TIMESHIP STEPHEN VALENTINE

To my parents, Frederick & Dora Valentine

Funding for the Timeship project has been generously provided by the Stasis Foundation, a nonprofit 501(c)(3) organization. I want to thank both the Foundation and its contributors for their continued support, as well as Saul Kent and William Faloon for inviting me to be a part of their vision.

Additionally, Timeship would not have been possible without the efforts of those who pioneered cryopreservation and life extension research, to whom we all owe a debt of gratitude.

info@timeship.org
www.timeship.org

TIMESHIP STEPHEN VALENTINE

THE ARCHITECTURE OF IMMORTALITY

EDITED BY

Stephen Valentine

PREFACE ESSAYS BY

Michael D. West, PhD
Robert A. Freitas Jr
Arthur L. Caplan, PhD

TEXT BY

John Lobell
Stephen Valentine

Content and Design
Stephen A. Valentine

USA Production Team
Jay Anderson
Agatha Cheuk
Ursula Damm
Mia Ihara
Jesse Kain
Jordan Rosenblum
Michael Wen-Sen Su

Copy Editing
Micaela Porta

Production
The Graphic Image Studio Pty Ltd. Mulgrave, Australia
www.tgis.com.au

Digital Production and Print
Leo Paper Products in Hong Kong/China

Published in Australia in 2009 by
The Images Publishing Group Pty Ltd
ABN 89 059 734 431
6 Bastow Place, Mulgrave, Victoria 3170, Australia
Tel: +61 3 9561 5544   Fax: +61 3 9561 4860
books@imagespublishing.com
www.imagespublishing.com

The Images Publishing Group Reference Number: 835

A Cataloguing-in-Publication record for this title is available from the National Library of Australia.

ISBN: 978 1 86470 324 5

TIMESHIP STEPHEN VALENTINE

THE ARCHITECTURE OF IMMORTALITY

STEPHEN A. VALENTINE

THE MISSION

Thousands of years ago, the first man discovered how to make fire. He was probably burned at the stake he had taught his brothers to light, but he left them a gift they had not conceived, and he lifted darkness off the earth.

Throughout the centuries, there were men who took first steps down new roads, armed with nothing but their own vision. The great creators—the thinkers, the artists, the scientists, the inventors—stood alone against the men of their time. Every new thought was opposed; every new invention was denounced. But the men of unborrowed vision went ahead. They fought, they suffered, and they paid. But they won.

– Howard Roark in The Fountainhead *by Ayn Rand*

IN CENTURIES PAST, families such as the Florentine Medici helped create the Renaissance through their patronage of art, architecture, and science. Today, as government and large foundations more often support the status quo, it is individual entrepreneurs and visionaries like Saul Kent and William Faloon who are the sources and supporters of new ideas—in their case, extreme life extension. Both of these extraordinary men are of that rare breed who do not always play it safe; they defy the odds and sometimes win. They are originals, true American pioneers with an unwavering commitment to their ideas and inspirations. They envisioned Timeship, and I created it.

Timeship's mission: to conquer aging and eventually death. Timeship will also be a scientific think tank and the world's center for the cryopreservation of the DNA of endangered species; human DNA, stem cells, and tissue samples for medical research; organs for transplant; and thousands of human patients traveling to the future.

The cryopreservation of humans, commonly known as cryonics, is the practice of cooling patients who can no longer be kept alive by contemporary medicine to temperatures so low that there is no tissue decomposition. They are then cryostored until reanimation may be possible in the future. This process is not currently reversible, and can only be performed after legal death. Cryonics is not a belief that the dead can be revived. Rather, cryonics is based on observations by neuroscientists that long-term memory, personality, and identity are stored by durable physical and molecular structures in the brain, not by transient electrical activity. Therefore, the thesis is that no one is really dead until their brain is completely destroyed and there is no chance of recovery, a condition known as "information-theoretic death."

We have until now accepted death as inevitable because of millennia of experience, but scientific advances suggest that aging and even death are genetically programmed and might be stopped. Today, scientists worldwide are engaged in extreme life extension research, and Timeship will be a state-of-the-art facility committed to this endeavor. The purpose of cryopreserving humans at Timeship is to enable those who might not be young enough to benefit from this research to get from the present to a future when indefinite life spans may become possible.

So why is death so bad? The award-winning documentary filmmaker, Errol Morris, asked Saul Kent this question. Kent's reply: "Because it is the end of life. The question is not even whether death is bad. Bad or good is not the issue: you're just not there. It is the disappearance of everything. Being dead to me means not *being*, forever."

The social, psychological, cultural, ethical, ecological, and religious implications of cryonics have been extensively debated. Socially, the question most often addressed is who can afford to be cryopreserved—the implication being that it is available only to the wealthy. In fact, just about anyone in an industrialized country can afford the procedure. Most of those already cryopreserved were of modest means, having financed the procedure through life insurance. There may be some ecological question as to how the planet will accommodate future populations, as well as those who were cryopreserved and brought back. This problem could be ameliorated by better ecological management of our resources and by the development of future technologies, including doing "more with less" as advocated by Buckminster Fuller. Does cryonics conflict with religious beliefs? Religious questions will have to be addressed by various faiths, but those involved in cryonics point out that most religions and cryonics are not mutually exclusive since cryonics is based on the sanctity of human life and the ethical imperative of continuing medical care for patients to help them live longer and healthier lives.

As the architect of Timeship, I realized from the beginning that the ideas behind this project were both extremely optimistic and dependent on advanced future technologies. I often wondered if these advances would ever be realized, or was Timeship just another Utopia—or perhaps even a "Udopia," a mad, impossible dream.

Imagine that the year is 1900. Who would have predicted the countless achievements and discoveries that would occur in what is actually just a "brief moment in time." In physics, we have seen relativity, quantum theory, and atomic energy. In engineering: giant dams, skyscrapers, and the interstate highway system. In mechanics: the automobile, the airplane, and the space shuttle. In electronics: television, personal computers, and the Internet. In medicine: antibiotics and organ transplants. And in agriculture: the green revolution. These and many other achievements might be summarized in just a few words: from "the man in the moon" to "man on the moon," an event that altered our perception of who we are in the universe, what we want to be, and where we are going.

Moving beyond the developments that carried us through the 20th century, we are at the beginning of two areas of technology that could have an even greater impact on our lives and that will be crucial to the future reanimation of Timeship's patients: genetic engineering and nanotechnology.

Genetic engineering holds the promise of curing diseases and vastly increasing food output from our shrinking planet. For some years now, much of our food has been genetically modified. Future advancements may even make it possible for us to feed ourselves without a dependence on animals. In the year 2000, we saw the "rough draft" mapping of the human genome, and now scientists are decoding the functions of each gene and studying their interactions in order to develop applications we can only begin to imagine. As this technology enables us to intervene at the level of DNA, many of the ills that afflict Timeship's patients will become curable, including perhaps even aging. And now some prominent scientists are predicting that the 21st century will become the "century of immortality," that children born today will probably live into their 100s, and that their grandchildren may never die as death is currently understood.

Nanotechnology envisions the creation of new materials and machines through molecular manufacturing by tiny robots, one atom at a time. It will make it possible to fabricate just about anything cheaply from the most common materials, and holds great promise for alleviating scarcity. Nanomedicine—medical procedures using nanotechnology—will be used for the repair of damaged cryonics patients. Patients who were cryopreserved using earlier methods will have considerable damage from ice crystals. Those who are now being cryopreserved with new vitrification methods will suffer far less damage, but both groups will need repairs during reanimation due to damage from the cryopreparation process, from their original diseases, and from aging.

Will such advances come about in time to help Timeship's patients within the next century? By their very nature, scientific breakthroughs are unpredictable, but what we can predict with confidence is the increasing rate of technological advances. The most famous example of this is "Moore's Law," which observes that the density of the computer chip doubles every eighteen months. We are beginning to realize that exponential growth is happening not only in computer chips, but in many technologies. We are on the verge of developments that will change everything about our culture. The term given for this is "the Singularity," meaning the point at which the growth curve is no longer inclining up, but is going almost straight up, a point we are fast approaching.

These technological developments may not only make possible the reanimation of Timeship's patients, but some researchers are now suggesting that they may also eventually answer the question of how the planet will support both them and future populations. Since the beginning, humans have evolved biologically, culturally, and technologically to deal with scarcity. Now, due to these emerging technologies, perhaps we might begin to contemplate a world of abundance.

When Saul Kent and I first discussed this project, he thought that the building should not represent any one religion or belief system, but rather represent all of humanity—a "faith in ourselves." He also suggested that the design be futuristic in its iconography, perhaps like a spaceship. However, a spaceship is a machine for traveling over distance, while we wanted a machine for traveling over time—in essence, a time ship. Thus, the approach to the building you see in this book and the name for the project: Timeship.

Ten years prior to Timeship, I worked at I.M. Pei's office as an architect on James Ingo Freed's design for the United States Holocaust Memorial Museum in Washington, D.C. There I had absorbed the idea of a building expressing its mission by symbolically embodying metaphors of memory. With Timeship, I realized that the building should symbolically evoke universal humanism and timelessness. My goal would be to endow the project with a symbolic language that speaks not of any one culture, religion, or philosophy, but of the transcendental qualities of life that are common to us all. It is the essence of symbols that they communicate what we may not be able to articulate directly, but that we know intuitively. They reflect human attempts to understand our surroundings and experiences, and thus our place within the cosmic order.

There may be no more important philosophical concept for Timeship to embody than permanence. Buildings begin to degrade as soon as their construction begins. Our society has an almost ingrained belief in impermanence, transience, and replacement. Timeship must, by virtue of its location and design, counter these values and convey in their place a sense of continuity and permanence—the values that will predominate in our culture once we can extend our lives into an indeterminate future. Architecture must symbolically convey the meaning of the activities it contains. Timeship challenges our notions of life and death, and must embody that fact. The challenge was to create an "architecture of immortality."

In 1997 in New York, I began the initial concept design of Timeship. With a limited budget, I engaged several architectural interns to help in the preliminary research, and had them explore concepts of death and immortality in various cultures. We looked at memorials, such as the Taj Mahal and the pyramids. We also investigated forts, bunkers, and castles because they were secure structures. I wanted to be bombarded with information—completely saturated—during the limited time I had. After ten weeks of total immersion and intense around-the-clock efforts, I finished the initial design and the concept drawings and models, which I presented to my clients at a life extension conference. Shortly thereafter, they concluded that further research on advanced cryopreservation methods would be needed before we proceeded to the next stage of the project. We now had an architecture of immortality, but the science had to catch up. Further development of the project was put on hold.

For decades, cryobiology was limited to using freezing as a method to preserve single cells or small tissues at liquid nitrogen temperature. In the early 1980s, a new method called "vitrification" was proposed for low-temperature storage of larger tissues and organs. With vitrification, biological material is brought to a glass-like or amorphous solid state without the formation of damaging ice crystals. Around the year 2000, breakthroughs were made with ice blockers and lower toxicity cryoprotectant solutions that began to fulfill the promise of cryopreserving large tissues. In 2005, the company 21st Century Medicine announced that it had removed a mammal's kidney and then vitrified it at a temperature of -130 degrees Celsius. Later that same day, it was rewarmed and successfully reimplanted into the animal, where it resumed normal functioning as the animal's only kidney. The vitrification and cryogenic storage of human organs for transplant had now become an increasing possibility for the future. In the cryopreservation of the human brain, vitrification is thought to be necessary to prevent destruction of tissue and therefore the information encoded in the brain. With this initial success of vitrification research, it was decided to move forward with Timeship.

To be certain that I had adequately defined Timeship's demanding design criteria, I assembled a group of thirty leading engineers, architects, and other consultants for a three-day conference, and posed the questions regarding the original concept design: what works, and what doesn't work? During the following twelve months, I continually met with design and engineering specialists as I prepared one of the most comprehensive program reports ever done for a building. The report's eight volumes established the ground rules for the further design of Timeship, including defining the criteria for selecting the site. I next began to assemble the engineering team for Timeship, including prominent firms with expertise in structural, mechanical, security, blast, and cryogenics design.

Besides being the foremost laboratory for cutting-edge, anti-aging research, Timeship will be the "Fort Knox" for the storage of precious biological materials. As such, it must be in a location that assures uninterrupted access and security against man-made and natural threats. In addition, all of its security, mechanical, and cooling systems must be absolutely reliable, designed with multiple backups, and able to be self-sufficient in the event outside sources of power are interrupted. To assure reliability, the cooling for the patients has been designed to work with few moving parts and minimal electricity, relying instead on the evaporative cooling of liquid nitrogen. In other words, these systems will work even if the electricity is cut off for several months. In addition, Timeship must keep its precious contents safe for possibly hundreds of years, so the building and its systems must have the highest level of security. In response to this requirement I adopted concepts from fortifications, such as Medieval castles, that have stood the test of time. Only then did we take advantage of current technological advances in security. As a result, we will have a proven ancient engineering approach enhanced with current and future technologies.

It will be some time before Timeship is fully operational, and we wanted patients, organs, human ova, and DNA of endangered species being cryopreserved in the meantime to benefit from the new vitrification technology. In 2003, we completed the conceptual design of a "temperature controlled volume" (TCV), a small-scale cryogenics storage container with Timeship features. In 2006, we received a patent for the TCV design.

The opportunities and challenges of the third millennium are only just becoming apparent. We are now beginning to realize that the exponential growth of technological development will soon create a world that we, at the beginning of the 21st century, will hardly recognize. Practical immortality may be just the beginning of this future to which Timeship is poised to take us. We have never thought of designing our society for indefinite life spans, so perhaps the question we should ask is, "Do we design the future or does the future design us?"

MICHAEL D. WEST, PhD

IMMORTALITY: SCIENCE AND SYMBOL

WHAT CAN SURPASS THE PYRAMIDS of Egypt as symbols of mankind's determined reach for immortality? Hand-carved in stone, the pyramids stand as magnificent monuments to mankind's love of life. What a brave people to raise such edifices toward the heavens—what warm hearts to hope their project would rescue their Pharaoh from the dust of death and transport him to a world where he would one day walk the sands of eternity.

The intellectual foundation for the ancient Egyptian religion was the concept of immortal renewal. In the realm of biology, the Egyptians noted that while the life of the individual animal or plant may be finite, and that while individuals may age and die, life itself is renewed eternally in the process of sexual reproduction. And similarly in the realm of astronomy, they noted that while the sun may appear to set in the west every day, it rises in the east the following morning in an eternal fugue—a testament (so they reasoned) of the eternal providence of the immortal gods.

So the ancient Egyptians asked themselves: What is the key to a human acquiring that immortality and traveling that path instead of merely facing the dissolution of the grave? Many ancient peoples thought that the problem was solvable. They described the problem of death as a "mystery," the solution to which the gods have given us clues in nature. These clues supposedly hinted at the path through the darkness of death into the light of eternity. They believed there were signposts to lead us through the labyrinth. The way out was to follow the immortal thread of life.

Shifting our gaze to the modern world, we find biologists on a similar quest. Again, in an effort to learn lessons from the immortal renewal of life, scientists have discovered new potentially life-saving therapies based on the actual cellular and molecular clockwork of immortality. The cellular basis of immortality is now known to reside in a thread-like lineage of cells that continually makes new babies that are born young. This lineage of cells is called the germ line. While somatic cells are "programmed" to necessarily age and die, germ line cells hold the potential for immortality. Looking back in time, we can see that they represent a continuum that has connected all of the generations of life on this planet for billions of years. And for the first time in history we have captured these immortal cells known as human embryonic stem cells in the laboratory dish. More astonishingly, we now even understand the molecular basis of this immortality. We have decoded the mystery of the mechanism of the immortality of these cells—a protein known as telomerase.

Telomerase is an immortality enzyme. Like the mythological Fates that spun the thread of life and set the life span of each mortal human, telomerase spins the nucleotides on the ends of the strands of human DNA that determine the limited replicative life span of human somatic cells. On the other hand, the germ line cells, like embryonic stem cells—the cells that connect the generations and that have been alive and replicating for millions of years—continually express the telomerase protein and are immortal.

Few areas of science and biotechnology have generated as many sparks as has human embryonic stem cell research and the use of nuclear transfer in medicine. The controversy is largely a symptom of science moving faster than political and religious leaders can accept. So the scientists working in this field have faced the challenge of not only unraveling the mysteries of the biology, but also of laboring to explain its promise to society. These scientists must try to explain that this new knowledge is benign and not malignant, that this technology is in the flow of what is noble in human history—that it is "pro-life" and not a "culture of death." Indeed, one could ask, what could be more pro-life than trying to banish death?

The challenge of modern science is to translate these discoveries into therapies that will radically transform medicine. This new venture has been dubbed regenerative medicine and it represents an effort to bring all of these and other important new technologies together so that one day we may be able to transport a cell from a human back in time to make young embryonic stem cells, and then from these immortal cells make young cells or tissues of any kind in the human body. Imagine taking the new and embryonic blood, vascular, skin, eye, and other cells that were present when you were formed in the womb, and using them to regenerate functions lost as a result of age. This may seem too futuristic, but let me suggest some concrete examples that may occur in my generation. We may soon be able to take cells from a patient, transport them back in time to young embryonic stem cells, differentiate them into primitive blood stem cells, and release them into the bloodstream to repopulate the blood with young immune cells. Or we might introduce into the blood cloned young endothelial precursor cells that can then spread throughout the vascular tree to "re-plumb" the cardiovascular system with young, healthy cells.

Embryonic stem cells hold the potential to generate a nearly infinite supply of young cells to regenerate tissues ravaged by age and degenerative disease. In the future, I expect medicine will increasingly utilize the natural immortality and regenerative potential of the germ line to save human life. In addition, with our understanding of DNA—the blueprint of human life—we can now combine the power of these immortal cells with our knowledge of the genome, enabling us to use our creativity in genetically engineering these cells to improve human health in ways we cannot even comprehend today. Embryonic stem cells and our growing knowledge of genetics are breathtaking in their potential to benefit mankind. But as exciting and new as these discoveries are, the ideology behind them is as old as civilization itself — to solve the mysteries of nature and to use that knowledge to save the lives of those we love.

When building the pyramids, the Egyptian engineers constructed only one pathway that led from the tomb to the heavens. Just one part of the eternally cycling sky seen in the heavens does not rise and set. The axis of the earth is directed into the region of the heavens close to the star Polaris. As a result, this part of the heavens remains constant in the midst of the cycle of the stars. It is as eternal as the heavens themselves. It was toward this part of the sky alone that the ancient Egyptians built a passageway to allow the spirit of the Pharaoh to escape to eternity. In the same way, the pathway set by modern regenerative medicine is the biology of the constancy of life in the immortal germ line through which we hope to engineer technologies that may lead, humbly and modestly, step by step toward that goal of human immortality.

What can surpass the pyramids of Egypt as a symbol of the strength of mankind's reach for immortality? Today, Timeship is founded on the bedrock of science rather than merely the symbols observed in the natural world. Ultimately, history may judge the impact of Timeship as more profound than that of the pyramids on the ultimate fate of the human species. In any event, Timeship symbolizes the belief of the laborers in the field of regenerative medicine that the individual human life is of infinite value and the project to save that life is worthy of our greatest efforts.

ROBERT A. FREITAS JR

NANOMEDICINE

THE ONLY IMPORTANT DIFFERENCE between the carbon atoms in a plain lump of coal and the carbon atoms in a stunning crystal of diamond is their molecular arrangement relative to each other. Currently envisioned, future technology will allow us to rearrange all atoms exactly the way we want them, consistent with natural laws, thus permitting the manufacture of artificial objects of surpassing beauty and strength that are far more valuable than diamonds. This is the essence of nanotechnology: the control of the composition and structure of matter at the atomic level. The prefix "nano" refers to the scale of these constructions. A nano-meter is one-billionth of a meter, the width of about five carbon atoms nestled side by side.

In coming decades, nanotechnologists will build nanoscale molecular parts like gears, bearings, and ratchets. Each nanopart may comprise a few thousand precisely placed atoms. These mechanical nanoparts will then be assembled into larger working machines such as nanopumps, nanocomputers, and even complete nanorobots. With medical nanorobots in hand, doctors will quickly cure most diseases that hobble and kill people today, rapidly repair most physical injuries our bodies can suffer, and vastly extend the human life span. This application of nanotechnology to the improvement of human health is called nanomedicine.

Microscale robots are already being investigated for *in vivo* medical use. In 2002, researchers at Tohoku University tested magnetically driven spinning screws intended to propel drug payloads through veins and into infected tissues, or even to burrow into tumors and destroy them with heat. In 2005, a team at the Swiss Federal Institute of Technology in Zurich fabricated a similarly powered microrobot small enough to be injected into the body through a syringe. The team hopes their device might be used to deliver drugs or to perform minimally invasive eye surgery. Moving still smaller in scale, experimentalists have used a rapidly vibrating micropipette to slice individual dendrites from single neurons without damaging cell viability. Other researchers have wielded tightly focused femtosecond lasers as nano-scissors to perform nanosurgery on individual chromosomes inside a live cell nucleus, and have dissected the cell wall of a single bacterium, layer by layer, using an atomic force microscope.

Even individual atoms can be manipulated. IBM scientists first demonstrated this in 1989 by using a scanning tunneling microscope to precisely position thirty-five xenon atoms on a nickel surface to spell out the corporate logo "IBM." By 2005, continuing progress in nanotechnology enabled Rice University researchers to build a molecular "nanocar" consisting of a chassis, two straight axles, and four wheels. The axles were rod-shaped acetylene-like molecules and the wheels were buckyballs, spheres of pure carbon containing sixty atoms apiece. The nanocar rolled around freely on a heated gold metal surface and was less than four nanometers across, a bit wider than a double strand of DNA.

Once we can build sophisticated medical nanorobots, how will they improve human health? Performance improvements up to 1000-fold over natural biological systems with similar function appear possible. For example, the respirocyte is a design for an artificial mechanical red blood cell. Natural red blood cells carry oxygen to our tissues and remove carbon dioxide. Respirocytes would too, but are made of much stronger diamondoid materials, not floppy lipids and proteins as in the case of living cells. This means the respiratory gases can be safely stored within the respirocyte at tremendous pressures—up to 1000 atmospheres—and can be loaded or unloaded, molecule by molecule, using mechanical pumps on the device's surface. This simple nanorobot is regulated by onboard computers, powered by glucose fuel cells, and controlled by a physician who communicates with the device via ultrasound signals beamed into the body from outside. A therapeutic 5-cc injection of respirocytes, just 1/1000th of total blood volume, duplicates the oxygen-carrying ability of the entire human blood mass and could, for instance, instantly revive emergency victims of carbon monoxide poisoning at the scene of a fire. Respirocytes would be just one micron in diameter, about the width of a bacterium and only 1/100th the volume of a natural red cell.

Another nanorobot design describes artificial mechanical white cells called microbivores that would seek and digest harmful pathogens including bacteria, viruses, or fungi. The pathogens are completely digested into harmless sugars, amino acids, and the like, the only effluents from this three micron nanorobot. Microbivores would completely clear even the most severe septicemic infections in hours or less, then be removed from the body. This is 1000 times faster than the weeks or months often needed for antibiotic-based cures. Related medical nanorobots with enhanced tissue mobility could similarly consume tumor cells with unmatched speed and surgical precision, eliminating cancer. Other nanorobots could be programmed to remove circulatory obstructions in just minutes, quickly rescuing even the most compromised stroke victim from near-certain brain damage.

Nanorobots could also be used for repair. For example, the vasculocyte would be a three-micron, 400-billion atom device that walks the inside surface of blood vessels atop telescoping appendages arranged on its underbelly. Future surgeons could employ these nanomachines to rapidly inspect and recondition the entire human vascular tree, virtually eliminating all cardiovascular disease. Somewhat larger 2.7-trillion atom nanorobots called chromallocytes would use chromosome replacement therapy to replace our old worn-out genes with new digitally corrected chromosome copies installed in every tissue cell of the body. This would eliminate genetic diseases and help reverse other cellular defects that lead to aging.

The worst medical insult is a fatal illness or trauma that necessitates a cold cryonic suspension. Here again, nanorobots can help the body heal itself because nanodevices can be designed to work effortlessly in extreme cold; even at absolute zero temperature, atoms display some vibrational motion. In one reanimation scenario, nanorobots tunnel ice channels throughout the frozen body, allowing precise mapping of the preserved physical structure to sub-micron accuracy. Such information, combined with genomic data and the patient's detailed medical history, would enable future medical technologists to design and execute a customized reconstruction plan. This work is analogous to that done by art restoration experts as they systematically renovate classic frescoes, like those found in the Sistine Chapel, to their original beauty and grandeur. Both the original illness and any damage created by the cold preservation process can be repaired. But absolute perfection is not required. We need only mend the body far enough that its own natural powers of healing and maintenance can take over, allowing the patient to resume normal life in his original state of good health without any further need for artificial support.

Many people over fifty believe that youth is wasted on the young. It is seldom appreciated that the wisdom of age is also wasted on the aged. That's because our old die too young, robbing the rest of us of the treasure of their accumulated centuries of experience. Timeship will transport these fallen to a future nanomedical era when today's diseases are historical curiosities, aging holds no terror for the human heart, and man's spirit, unshackled at last from a ticking biological clock, is freed to seek its ultimate destiny in the universe.

ARTHUR L. CAPLAN, PhD

THE ETHICS OF EXTREME LIFE EXTENSION

BIOMEDICAL RESEARCH UNDERWAY today holds the promise of extreme life extension. Not just a prolonging of the end of life by five or ten years, but the possibility of extending the average healthy life span to three or four times what it is now. However, not everyone thinks that this sort of extreme life extension is a good idea. Critics argue that a longer life is not necessarily a better life and that the world could become crowded with doddering, demented, and demanding old people. Others add that it is objectionable on religious grounds for us to violate God's will and reach for immortality, and still others argue that we may disrupt evolution. These people feel that lengthening life is "unnatural," that using medicine and engineering to extend our lives beyond the proverbial three score years and ten is to vainly and arrogantly interfere with an essential, defining feature of humankind—mortality.

Are these critics right? This is a question now being posed to bioethicists. So first, some thoughts on bioethics. The role of bioethics is not so much to tell us in absolute terms what is right and wrong, but rather to provide frameworks within which all of us can democratically discuss issues and contribute to informed decisions. In establishing such frameworks, we need to address four areas of concern, the first being the individual. What will be the effects of a proposed new procedure or technology on the person who will be subjected to it? What are the possible risks and potential benefits and how should they be weighted against each other? If extreme life extension is to be achieved through such technologies as radical gene therapy or invasive nanomedicine, the procedures will have to be well tested in advance of use on humans, and those who chose to undergo them will need to be aware of the risks.

The second area of concern is society. What will be the effect of a proposed new procedure or technology on the larger community? What will it cost, who will pay for it, and what about those who cannot afford access? Extreme life extension will have major impacts on all of our social structures, and the full range of its implications must be broadly discussed by all interested groups before research is permitted or services are offered.

The third area of concern is the law. Is a proposed new procedure or technology legal under existing law? Should the law be changed to address new circumstances? If extreme life extension proves feasible it will require legal changes in many areas, from the regulation of biomedicine to the handling of estates and trusts.

The fourth and perhaps most important area of concern is the culture. How will a proposed new procedure or technology affect existing mores, social conventions, and religious beliefs? How will it relate to the histories and traditions of those involved? Here extreme life extension poses the most concerns, for it challenges thousands of years of traditions and will potentially change our understandings of who we are at the deepest levels.

We can dismiss concerns about immortality. Given the frailty of our bodies and the realities of living in a physical world, immortality is probably beyond the reach of science and medicine, period. But what about the ethics of seeking the secrets of living two, three, or four times as long as the oldest amongst us today? Will such longer life spans mean more social misery, economic havoc, and environmental stress? One way to approach this question is to see how humanity has adjusted to much longer lives in the past. We have seen roughly a doubling of the average life span since the times of the ancient Assyrians, Hebrews, Greeks, and Romans. And while the world has experienced much turmoil since those times, the social order has been able to adjust again and again to this and other changes brought about by advances in public health, medicine, nutrition, and science. Of course, increased population has led to problems including environmental damage, but technological advances have, on the whole, greatly improved the quality of life. Few people in modern Syria, Israel, Greece, and Italy would trade their lives today for the much shorter lives lived by their ancestors thousands of years ago.

In order to cope with longer lives we will need new social structures, but we have already seen such adjustments as retirement, leisure, adult education, second careers, and so on. Extreme life extension will, of course, call for even more imaginative innovations. We will need policies that ensure that a fair proportion of resources is devoted to the young, that seniority on the job does not become stasis in the workplace, and that we do not use medical technology overly aggressively once life has become too painful to endure. But such policies will surely be developed, and society will most probably adjust in ways that make longer lives enjoyable, productive, and meaningful.

The prospect of extreme life extension also raises religious questions for some. In the Bible, God causes humans to age and die as a punishment for the sins of our ancestors in the Garden of Eden. But extreme life extension is not immortality, and there were many figures in the Bible who lived far beyond even the life spans being discussed here. Religion, while wary of change, is also open to the idea that man can and should seek to improve his lot and that of his children.

Others question whether extreme life extension poses a threat to the natural order of evolution. From this perspective, the function of aging is to ensure the death of organisms in order to allow evolutionary change and new adaptations. But we do not need to die at seventy-five to get out of the way of the young, since natural selection does not care how long we live once we have stopped reproducing.

Then there is the issue of self-interest, with some saying that it is selfish to want to live for hundreds of years. But why is that any more selfish than it is to seek medical care for a failing heart or to watch one's diet or level of stress in order to live a longer, better life? Self-regard is not the same as selfishness. Those who want to live longer are not necessarily self-obsessed. They may want to see their great great grandchildren, contribute more to society, help advance the common good, or try to provide help to their family and friends for as long as they can. Not every reason to live longer is selfish.

Finally, there is the issue of priorities. Given all of the problems in the world today, is extreme life extension something on which we should expend money and focus scientific research? However we might turn the question around and ask, given all of the silly things on which we expend resources, why should life extension come under criticism? Especially since life extension research will surely produce benefits and relieve pain and suffering even for those who choose not to take advantage of living longer.

A thousand years ago, when just a few could read, we would not have predicted that eventually most people would read and have access to symbolic communications that transcend place and time. Five hundred years ago, when only Renaissance nobles had liberal educations, we would not have predicted that most people would eventually swim in an encyclopedic awareness of the workings of the universe from the atom to the brain to the galaxies. In this light, we might think of extreme life extension as presenting the possibility of bringing into expression human potentials that we all possess, that we all deserve to experience, but that we today do not yet even know we have.

01 ORIGINS OF TIMESHIP

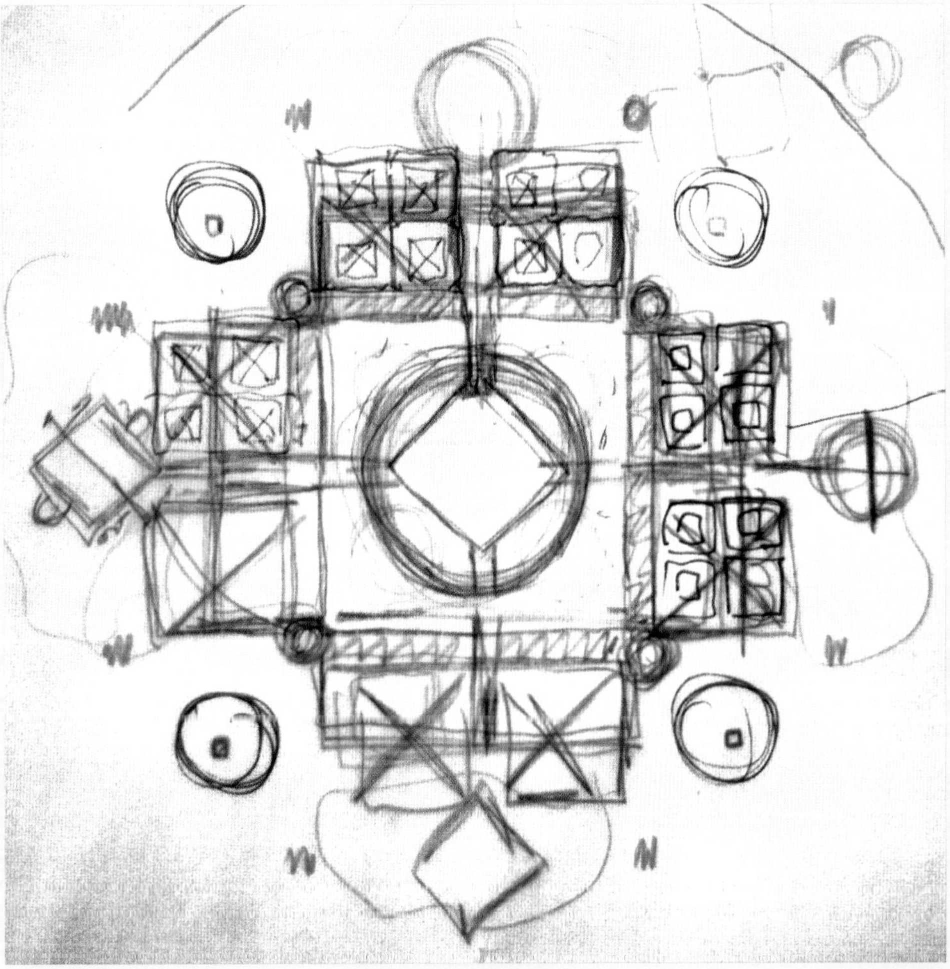

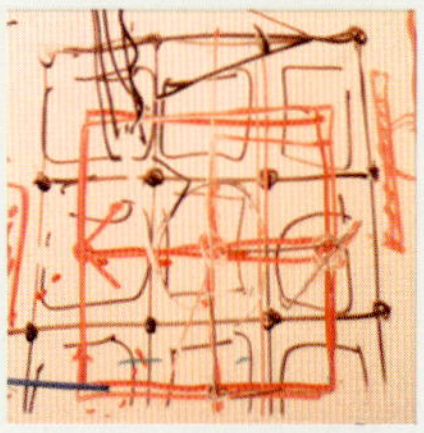

ORIGINS OF TIMESHIP

WE HAVE LONG BEEN FASCINATED by immortality, as we see for example in the Greek myths about immortal gods and mortal humans who are occasionally granted immortality. These tales also contain cautions, as in the story of the Cumean Sibyl who won immortality from Apollo, but neglected to stipulate that she would not age. However, for much of human history, despite thinking about immortality, we have been resigned to aging and death. As with the weather, there was little we could do about it.

Metaphysical immortality, the notion that an entity could never die, is outside the realm of science. But developments in science, particularly genetics, may eventually make possible extreme life extension of hundreds of years, and perhaps even practical immortality, sometimes called technological immortality, in which the body does not age, and one would die only from trauma such as an accident. Indeed it now appears that aging and death are programmed into our genes and that the new frontiers opened by DNA research may make it possible to identify the genes that cause aging and turn them off.

A number of scientists in the field estimate that extreme life extension may be twenty to a hundred years off. What does that mean for people today who are in their fifties or older? How might they get from this present of aging and mortality to a future of indefinite youthfulness? One answer is time travel, not in the science-fiction sense, but through cryopreservation—be cryopreserved now and reanimated after decades or centuries when one can be "reconditioned" to a healthy youthful life that will last indefinitely.

In 1964, the physics professor Robert Ettinger published the book, *The Prospect of Immortality*, proposing that a person could be frozen immediately after legal death and reanimated at a future time when the cause of death could be cured and advanced technologies could restore the person to youthful health and could halt aging. The cryonics movement grew out of these proposals, and in 1967, the same year as the first heart transplant, Dr. James Bedford became the first person to undergo cryonic suspension. He was placed in a capsule that was developed using super insulation and high vacuum cryonic technologies that had recently been developed for the upper stages of the Saturn rocket that eventually took us to the moon. Today over one hundred people are cryopreserved at liquid nitrogen temperatures, in which state they can remain indefinitely with no deterioration. Additionally, over a thousand people are signed up for the procedure at their deaths. Reanimation of those who have been cryopreserved is not possible today, but the increasing rate of biotech and medical advances suggests that it will be someday.

It is for the cryonic storage of patients, as well as to be a center for life extension research, that Timeship was conceived. Timeship will be the world's most secure and technologically advanced facility for the long-term storage of cryopreserved biological materials. Timeship will also house the most advanced laboratories doing research on every aspect of life extension, from nutritional supplements that may slow aging now, to research at the genetic level that could eventually stop and reverse aging. This research will also include all aspects of the cryopreservation process—not only the preparation of patients traveling to the future and their reanimation, but also the preparation and storage of human organs for transplantation.

At this time most organs cannot be frozen due to the damage caused by ice crystals, so genetically matching organs must be obtained immediately after a donor dies. The vitrification process being developed for Timeship may make it possible to create banks of vitrified organs and thereby save the lives of many who today cannot find matching organs in time. This research will also include nanotechnology, which will be important for the creation of miniature robots, or "nanorobots," that will enter the body to repair damage from disease and the cryo-preservation process.

Thus Timeship's mission is in two parts: life extension research and cryo-preservation of patients and biological materials. To fulfill these two missions, Timeship's six-acre structure is designed to provide safety and security against multiple threats, both natural and human-made, including terrorist attack, sea level changes due to global warming, and interruption of energy supplies. Timeship is designed to remain operational for hundreds of years.

The exterior and interior of Timeship echo themes of security and rebirth both functionally and symbolically. The architect of Timeship, Stephen Valentine, felt that symbolism was important in the design, remarking that "architectural symbolism is a primary way for us to absorb the meanings and implications of radical change, and Timeship has the potential to change our notion of death, and thereby our notion of human life itself." Furthermore, Valentine felt that since Timeship would need to be operational for hundreds of years, it was important that its design be timeless and not just reiterate the fashions of the moment.

Why might someone want to live indefinitely? In addressing this question, we have to realize that the very understanding of who we are as human beings has changed over time. Primitive peoples saw themselves as one with nature, and regarded death as a melting back into the earth from which they had come. Early high civilizations, such as the ancient Egyptians, saw the beehive-like social group and its relationship to the order of the cosmos as primary, not the life and death of the individual. It is in the West that we see the emergence of the individual.

Today we invest in individuals, protecting and nurturing them, and providing them with educations that will enable them to become autonomous, self-actualized people, capable of pursuing their individual paths in life. We see each individual as having a unique view of the world, unique talents, and an irreplaceable "self."

Until now we have experienced our lives as having a natural arc, with parts devoted to youth, maturity, and old age, each with their functions and desires. But recently we have begun to challenge those boundaries. For example, the function of learning, once primarily confined to youth, has spilled over into maturity and even old age, and just as we begin to fully master our disciplines we find ourselves aging. In addition, in the past one's body might have been accumulating aches and pains by the age of sixty, provoking a feeling that the time for rest might be approaching. But today many youthful sixty- and seventy-year-olds are running marathons.

Some are beginning to ask, what is the point, after all of this effort to develop one's unique individual self, of living for only a few decades? Why not further develop one's talents over centuries, and give society the opportunity to benefit from contributions we cannot yet imagine? Extreme life extension might lead to a "long view." Rather than think in terms of humankind eventually journeying to the stars, we might think in terms of someday journeying to the stars ourselves. We might not master just a narrow discipline, but a broad range of disciplines, and accomplish scientific or artistic achievements not previously possible. We might not just plant and eventually harvest vineyards, but also olive groves. And we might not just plan for the educations of our children, but also of our great-great grandchildren.

The human race survived for millennia without the notion of the individual. But it was only with the liberation of the individual in eighteenth-century Europe and America that we saw the explosive developments of science and modern culture. Today the average person commands more wealth, knowledge, and opportunity than did the princes of the ancient world. The individuality that exists in us today existed as a potential in the past, but no one saw it. What further potentials lie in us, waiting to be released when we can extend our lives over hundreds of years? It is the vision of these potentials that motivates those who advocate extreme life extension, who spend their resources on the research to achieve it, and who are building Timeship.

figure [*a*]

THE IDEA OF SUSPENSION AND REANIMATION has long been with us, as we see in such stories as *Snow White, Sleeping Beauty,* and *Rip Van Winkle.* The desire to experience the future also has a long tradition, as we see in H. G. Wells' *The Time Machine.* What does the future hold? How can we live long enough to get there? Are we satisfied with our limited years or is there more we wish to accomplish?

When we review the scientific and medical advances of recent decades, we realize it is not unreasonable to expect that in as little as one hundred years, disease, aging, and even death may be conquered. But how do we get from the present to this future? Timeship is being built to care for cryopreserved patients, and provide laboratories for the regeneration, rejuvenation, and reanimation of humans.

[*a*] Benjamin Franklin said, "Nothing in life is certain except death and taxes." However, Franklin was a scientist with broad interests, and he did speculate about suspension for eventual reanimation.

[*b*] In 1964, physics professor Robert Ettinger wrote *The Prospect of Immortality.* His theory was that a person could be frozen immediately after legal death, with the prospect of eventual reanimation by future advanced medicine that could cure diseases, reverse aging, and repair the damage of freezing. The cryonics movement grew out of this book.

[*c*] We all know the story of Sleeping Beauty, who slept for 100 years before she was awakened by Prince Charming's kiss. We have a long tradition of various versions of this and related stories.

figure [*b*]

figure [*c*]

figure [*a*]

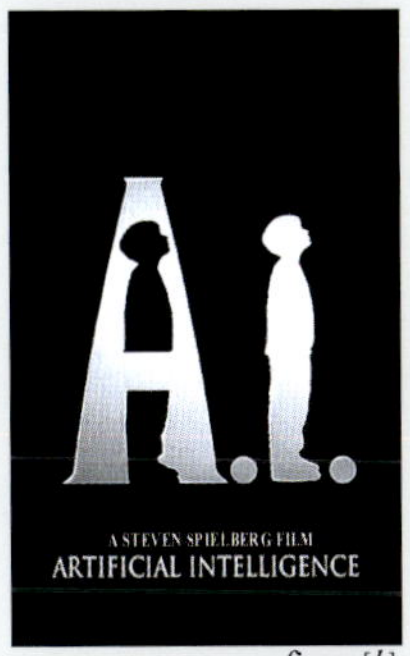

figure [*b*]

figure [*c*]

[*a*] Woody Allen's *Sleeper,* produced in 1973, included an imaginative portrayal of cryonics.

[*b*] In *A.I.,* a boy is cryopreserved until his injuries can be treated. He then comes into conflict with his robot replacement, raising questions about love and humanness.

[*c*] In *2001: A Space Odyssey,* the crew is suspended for part of its long journey to Jupiter.

CRYONIC SUSPENSION (human cryopreservation) may at first seem like a new and even extreme idea, perhaps suspect to much of the general public. But cryonic suspension has, in fact, been understood and well accepted in the public imagination for almost four decades. Consider the numerous popular movies in which it has been featured, from science fiction to comedy to romantic drama, including: *2001: A Space Odyssey, Sleeper, Forever Young, Late for Dinner, Batman and Robin, Demolition Man, Austin Powers, Artificial Intelligence: AI, Vanilla Sky,* and *Minority Report.*

Movies and television are often a means by which we are introduced to and debate new ideas that are challenging our culture. They give us the chance to examine the human and emotional impact of a new technology or a major change. As described above, cryonic suspension has been so addressed for many years, and now genetic engineering, cloning, climate change, and terrorism, all pertinent to Timeship, are being similarly addressed.

figure [*a*]

THERE IS A LONG TRADITION of exploring alternatives to human beings as we now exist. The ancient Egyptians took the notion of an afterlife literally enough to bury with the dead the things they would need in the next life. Mary Shelley explored the making of a new kind of being in *Frankenstein,* and there have been many science-fiction depictions that marry a disembodied brain to the latest technologies. Many scientific developments that were once controversial are now widely accepted and even commonplace. The Frankenstein image of the mad scientist as a threat now seems quaint. Similarly, genetic engineering was initially challenged, while today much of our food is genetically engineered.

New fields, variously called posthumanism, transhumanism, better-humans, etc. are hotbeds of discussions about what possible future human development might be like.

figure [*b*]

[*a*] An Egyptian mummy, once believed to be a way to travel to the next world, being excavated.

[*b*] Illustration for a science-fiction story about a disembodied head connected to the technology of the day.

[*c*] Dr. Frankenstein with his monster, once a scary image, is no longer threatening to most of us as we learn about recent advances in biotechnology.

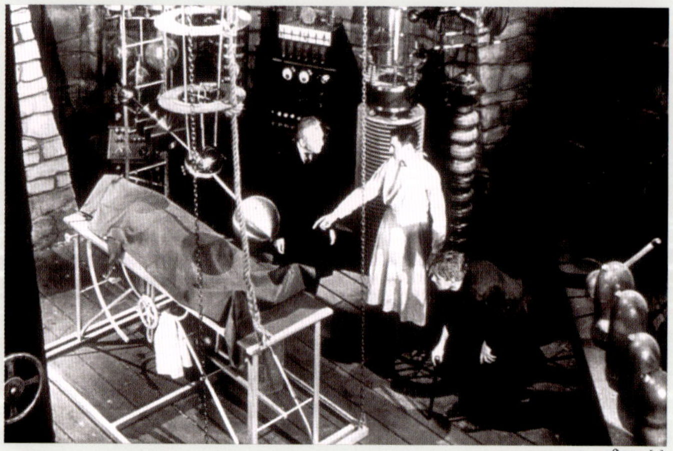

figure [*c*]

[*a*] Dr. Christiaan Barnard's first heart transplant patient died in 18 days, but immunosuppressant drugs now make the procedure common.

[*b*] Over a quarter of a million babies have been born through frozen embryo transfer.

[*c*] In January 1967, Dr. James Bedford was the first person to undergo cryonic suspension. Techniques have since advanced.

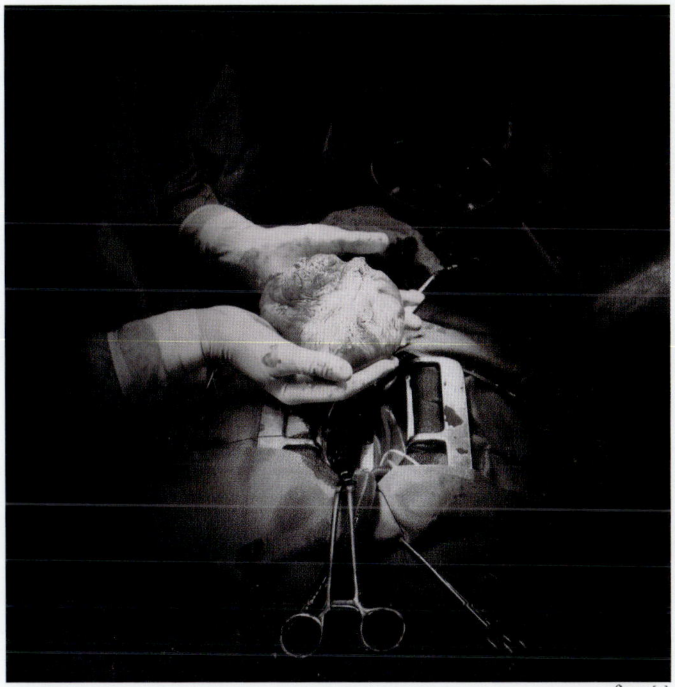

figure [*a*]

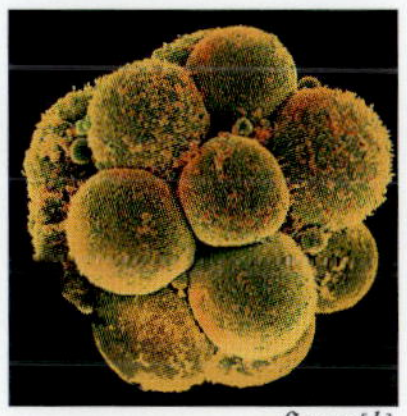

figure [*b*]

THE LATE 1960s was a period of social unrest and scientific and engineering achievements. The first heart transplant was in 1967 and the first cryonic suspension took place that same year. We landed on the moon two years later.

When Dr. Christiaan Barnard performed the first heart transplant, many people accused him of offending God, nature, and humanity, and some questioned whether the patient would be conflicted by two souls. Today, heart and other organ transplants are common, widely accepted, and save thousands of lives annually. They will become even more common as the technologies being developed for Timeship make it possible to create banks of cryopreserved cells, tissues, ova, and organs. Just as organ transplants have come a long way since the 1960s, so has cryopreservation. Recent developments in the field have now led to the design of Timeship.

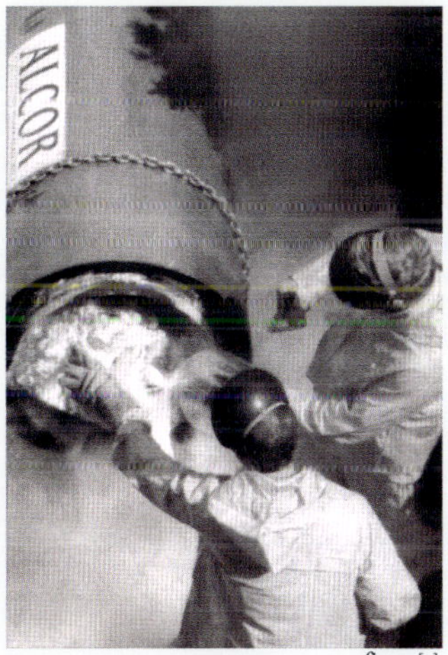

figure [*c*]

[*a*] & [*b*] A team at the Alcor Life Extension Foundation cryopreserving a patient using the most advanced procedures. The setting is similar to an operating room in a modern hospital, and much of the same equipment is used, such as a heart lung machine. Anticoagulants must be introduced into the patient's system so that circulation continues to all parts of the body. Cryoprotectants are introduced to prevent ice crystals from forming during cooling.

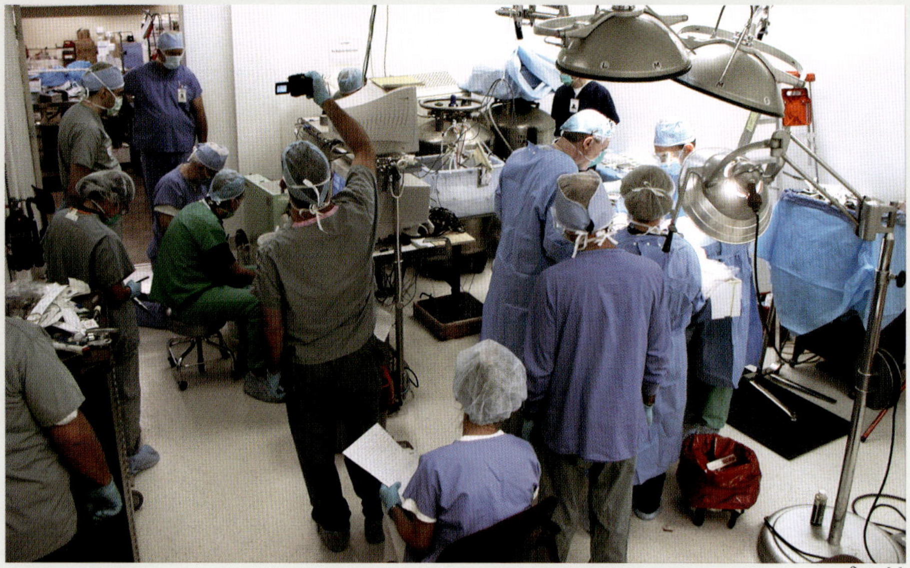

figure [*a*]

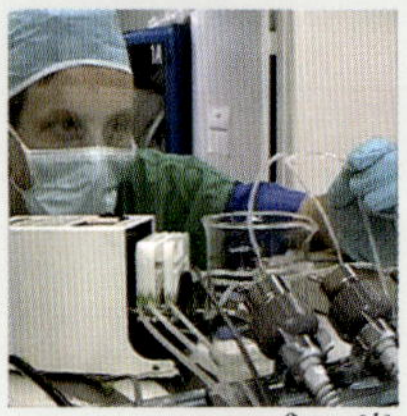

figure [*b*]

WHILE TIMESHIP REPRESENTS major advances in cryopreservation by using vitrification, the process has a long history. Since the first patient was suspended in 1967, more than one hundred people have been cryopreserved and more than one thousand have made legal and financial arrangements for the procedure, usually through life insurance. There are several organizations currently providing these services, which Timeship will bring to a new level. As the new technologies that will be used at Timeship are perfected and become widely known, there will be a rapid increase in demand for cryopreservation services.

Currently, cryopreservation begins as soon as possible after pronouncement of death. Under ideal circumstances, a special team is on hand waiting with its equipment. Blood circulation and breathing are artificially restored, a series of medications is administered, and the patient is rapidly cooled to protect the brain. The full procedure is as complex as modern major surgery, using much of the same technology, and requires a highly trained team. The cryopreservation procedures that will be used at Timeship will be based on the latest scientific and biomedical breakthroughs.

figure [*a*]

[*a*] Patients who have chosen to have just their brains (within their heads) preserved are placed in individual aluminum containers. Timeship will use next-generation organ pods for these patients.

[*b*] & [*c*] Currently whole-body patients are stored four to a stainless steel vacuum-walled dewar. The dewar is filled with liquid nitrogen (LN_2) and periodically topped off, assuring a constant, easy to maintain –196°C. The state-of-the-art vitrification process that will be used at Timeship requires a somewhat higher temperature.

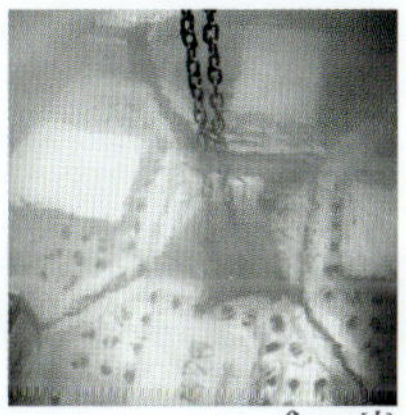

figure [*b*]

figure [*c*]

BIOLOGICAL SPECIMENS STORED in liquid nitrogen are kept in vessels called dewars. There is no chemical activity at the temperature of liquid nitrogen, and therefore no deterioration of the specimens. Dewars are constructed of double walls of stainless steel with a vacuum in between—they are essentially giant thermos bottles. Their operation is simple: liquid nitrogen (LN_2) is poured in, it evaporates and draws off heat, more is poured in, and the contents stay at a constant –196°C. Cryopreserved patients today are maintained in this manner.

The patients and biological specimens at Timeship will be cryopreserved using advanced methods to prevent damage from ice crystals, and must be maintained at around –140°C (instead of –196°C), necessitating complex technologies at the core of Timeship.

figure [*a*]

[*a*] One of a dozen breakout groups of consultants reviewing the initial design for Timeship, focusing on what works, what does not work, and what needs to be addressed next.

[*b*] The comprehensive eight-volume "Timeship Program Report," including: Site Analysis; Architectural Space Planning; Symbolism; Cryogenic and Cold Storage Systems; Security Design and Blast Design; Life Support Systems; and Energy.

[*c*] "Program Workshop Report" analyzing the strengths and weaknesses of the initial concept design and outlining strategies for the next stage of design development.

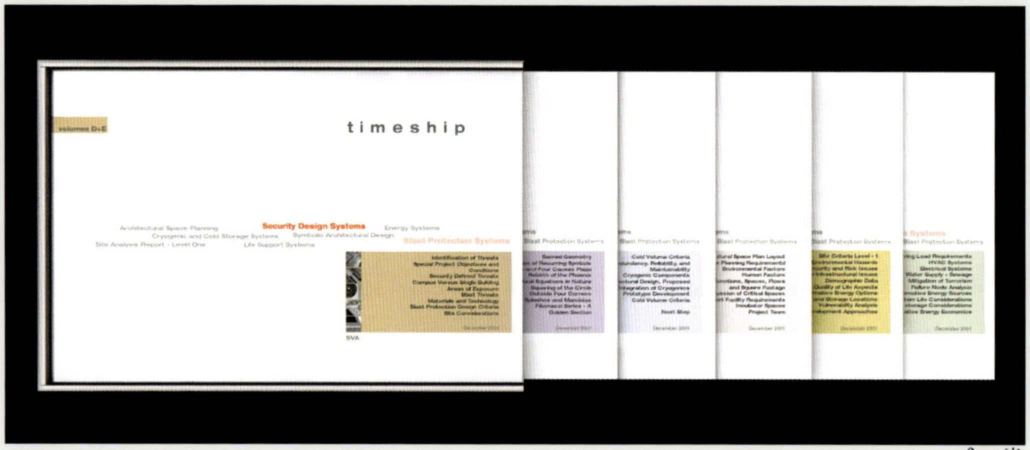

figure [*b*]

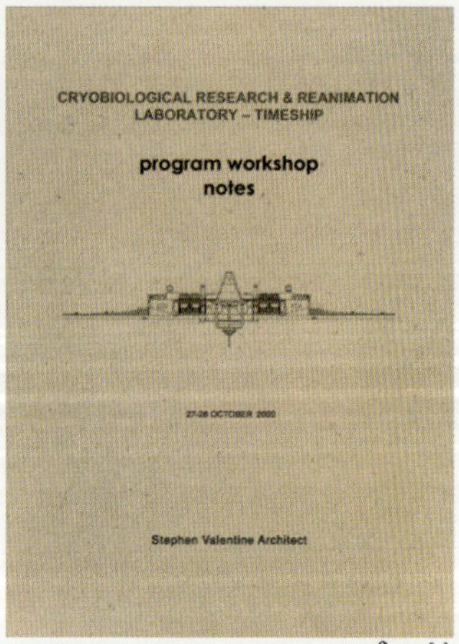

figure [*c*]

AFTER CREATING THE INITIAL DESIGN for Timeship, Valentine convened a group of thirty architects and engineers for three days to review the concept, ask what does and does not work, and produce a Program Workshop Report with strategies for the next stage.

Valentine then retained leading experts in site selection; structural, blast, and mechanical engineering; seismic design; cryogenic design; laboratory design; and security to produce the multivolume "Timeship Program Report," which described in great detail the requirements for the building. This is one of the most comprehensive program reports ever done for a building.

[*a*] The Municipal Art Society of New York mounted a major exhibition, "Timeship: The Architecture of Immortality," comprised of panels with detailed graphics and text; models, including a life-sized, whole-body pod; and video displays. The exhibition led to vigorous discussions and it will be displayed internationally.

[*b*] Concept plans at four levels of Timeship.

figure [*a*]

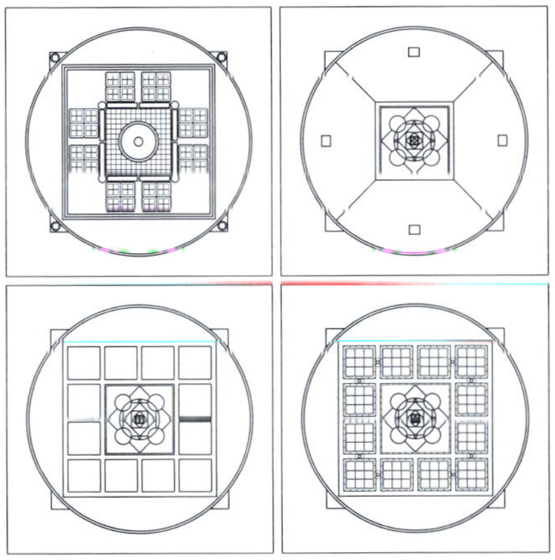

figure [*b*]

A MAJOR EXHIBITION OF TIMESHIP was held in the galleries of the Municipal Art Society of New York. It generated intense discussions about the scientific feasibility and the ethics of both cryopreservation and practical immortality. Timeship is a revolutionary project, at the cutting edge of a dozen fields in science and technology, with the potential to eventually bring about vast social and cultural changes, and even changes in our understanding of life and death.

Those in the field of cryonics have long engaged in discussions of the implications of these issues, and the advent of the Internet has greatly intensified these discussions. Now, as the Timeship exhibition travels worldwide and as the project moves forward, Timeship will become a major focus for discussions between cryonics advocates and others, including scientists, doctors, politicians, bioethicists and religious leaders.

02 LIFE EXTENSION RESEARCH

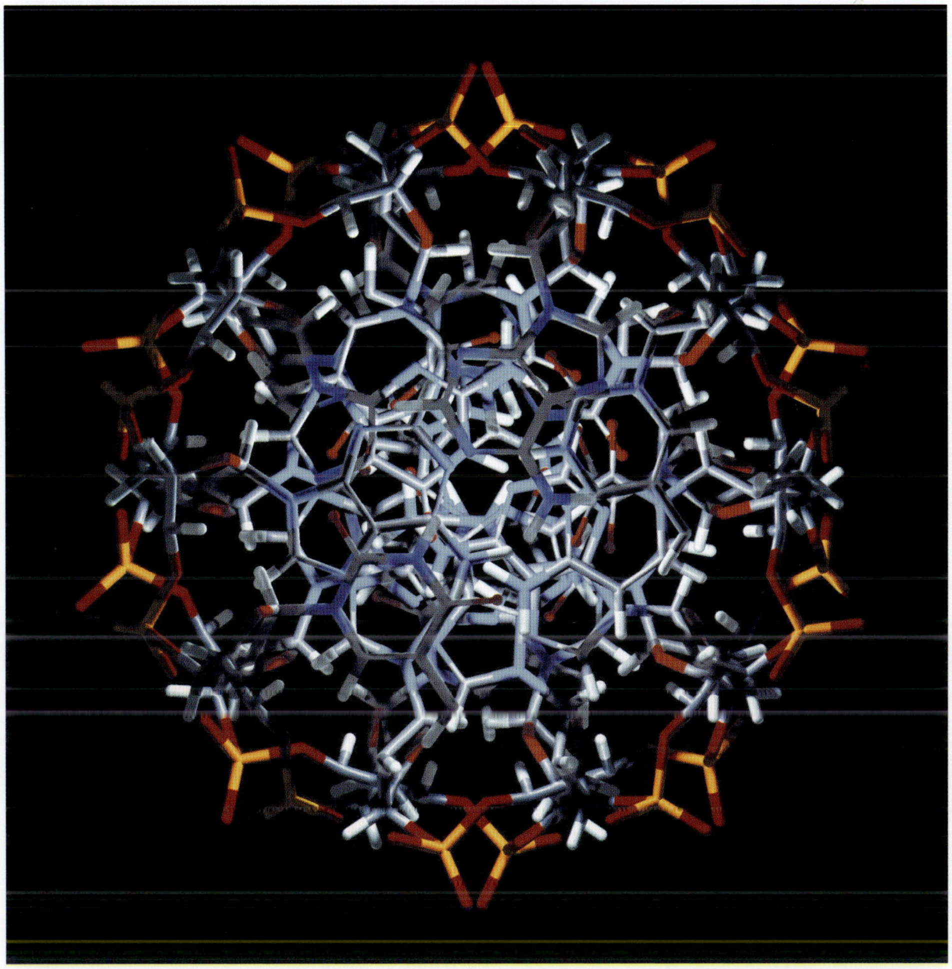

LIFE EXTENSION RESEARCH

NOTIONS OF THE EXTENSION OF LIFE are found throughout history, sometimes as a "fountain of youth." With the rise of modern science and medicine such notions are becoming real.

We can see progress in life extension in several stages. The first stage is the prevention and treatment of bodily wear-and-tear and of disease. Public health measures such as safe water, proper sanitation, clean air, good nutrition, and safety in the workplace have added more years to the average life span than all of medicine combined. Indeed, until the last hundred years medicine had not been very effective. However, with the development of vaccines and the widespread use of antibiotics after the Second World War, medicine also began to have a major effect on longevity. In the past forty years, aggressive medicine has added years to the average human lifespan.

The second stage of life extension involves the advanced repair of the body. Although little heralded, modern dentistry has done much to ameliorate suffering, and has greatly aided overall health and longevity. In the 1950s and 1960s, the first successful kidney and heart transplants took place, opening the door to a whole range of procedures from stents to open clogged arteries to transplantation of most of the major organs.

Organ transplants are now common, and will become more so with developing technologies. Dr. Gregory M. Fahy of 21st Century Medicine (21CM) has removed a rabbit kidney, vitrified it (cooled it to a solid state without the formation of damaging ice crystals), and successfully rewarmed and reimplanted it. With banks of vitrified human kidneys and other organs readily available, the number of people who die each year waiting for a matching kidney will be greatly reduced. Even that technology might soon become obsolete if a perfectly matched kidney can be grown from the patient's own stem cells, while other stem cell research holds great promise for a whole range of therapies.

However, while such procedures extend lives that would have ended sooner, they do not slow or halt the underlying aging process, which will be the third stage of life extension. There is hope that various nutritional practices, dietary supplements, and drugs might provide some help in this area. For example, it has been noted that reduced-calorie diets greatly extend the lives of laboratory animals, and it is conjectured that they might do the same for people. Since few people want to be hungry all the time, BioMarker Pharmaceuticals is seeking to identify the genetic process underlying this phenomenon, and to reproduce it with drugs or other therapies. Many nutritional supplements enhance health and some may slow the aging process. The Life Extension Foundation is sponsoring research on the roles of many nutrients, including antioxidants that prevent excessive free radical activity from damaging cells. There is also evidence that reducing levels of the protein insulin-like growth factor-1 will result in increased life span.

The fourth stage in life extension will come with understanding and reversing aging. It now appears that the number of times a cell can divide is programmed by telomeres at the ends of each DNA strand that shorten with each cell division. Extending the length of telomeres might prevent aging. Once the genetic causes of aging are identified, there is the possibility of altering the genes in parents' sperm and ovum before fertilization so that they give birth to offspring who will not age beyond maturity. It may even be possible to alter the genes of adults by introducing new DNA using viruses as vectors. The rapidly developing field of nanotechnology also holds great promise. Nanotechnology anticipates the eventual construction of nanorobots smaller than blood cells that will travel through a patient's system hunting down pathogens, repairing damaged cells, and perhaps even entering cell nuclei to reengineer DNA.

For some who speculate about distant possible futures, the fifth and perhaps final stage of life extension may come from advanced DNA research and from a possible merging of biology and electronic (or other future) technologies. DNA technology has already led to the creation of plants and animals that never existed before, and it may eventually lead not only to cures for major degenerative diseases, but also to human enhancements. As for electronics, we already interact with our computers by typing on the keyboard and looking at the screen. Many deaf people have cochlear chip implants allowing them to hear, and some blind people have chip implants in the optic centers of their brains allowing them to see. A few paralyzed people with implanted chips can control computers and mechanical arms just by thinking. Technological optimists foresee a time when we may have chips with entire libraries implanted in our brains and linked to our neurons. At the extreme, some foresee a time when we will be able to transfer aspects of ourselves into computers for backup, as insurance against neurological damage.

Timeship will be a major center for leading-edge life extension research, building on the pioneering work of the Life Extension Foundation, BioMarker Pharmaceuticals, 21st Century Medicine, and Critical Care Research. Cryopreservation research that will make it possible to store human organs for reimplantation, and reanimate them, is also a priority of 21st Century Medicine.

It is impossible to predict when and even if these developments will actually happen. However, the acceleration of technological progress over the past hundred years suggests that most of them, and many that we now are unable to predict, will happen sooner rather than later. But many technological optimists are aware that they may not live to see the breakthroughs that would assure their personal extreme life extension, and some see cryopreservation as a way to get from "now" to "then" in order to eventually be reanimated and live indefinitely.

While human cryopreservation has been around for over forty years, it is only in the past decade that new approaches have made it far more promising. Patients previously were frozen to the temperature of liquid nitrogen. At such low temperatures all biological and chemical activity stops and biological systems remain unchanged for hundreds of years. However, using this process leads to the formation of ice crystals, causing severe damage to cells. Today, new vitrification technology allows tissues to solidify without the formation of ice crystals. In 2006, a team led by 21CM reported almost 100 percent viability of vitrified mammalian brain slices (specifically, the brain center that encodes and decodes memories). In this case, viability was measured by the ability to actively maintain high intracellular potassium content and low intracellular sodium content at normal body temperatures, which is a prerequisite for brain electrical activity.

The technologies thus far described range from cleaner drinking water to organ transplants, all the way to having oneself downloaded into a computer and later uploaded into a new body. Although some people have a comfort limit as to how far they think these technologies should go, we should remember that when the first heart transplant was performed, many people condemned it. Today, organ transplants are common and there are few objections. As with most major technological advances, extreme life extension and possible practical immortality will be enthusiastically welcomed by some and objected to by others. Some regard aging as natural, something with which we should not interfere. Others regard aging as a killer disease that needs to be eliminated.

But the new technologies under discussion here raise deeper questions. What is the "self?" Is identity synonymous with the brain and the memories it stores? Are memories retained during the cryopreservation process? Would the self still be the same if only the brain were preserved and a new body were cloned around it? Would the self still be the same if the body were gone and the memories were transferred to electronic storage? And would we still be "human?" Would it matter?

All of these questions offer endless discourse for philosophers, ethicists, and theologians who have long discussed the meaning of our mortal human lives. In the future, perhaps they will have to address the meaning of practical immortality: lives that have no natural end other than by accident or choice, perhaps even existing on the future equivalent of computer chips.

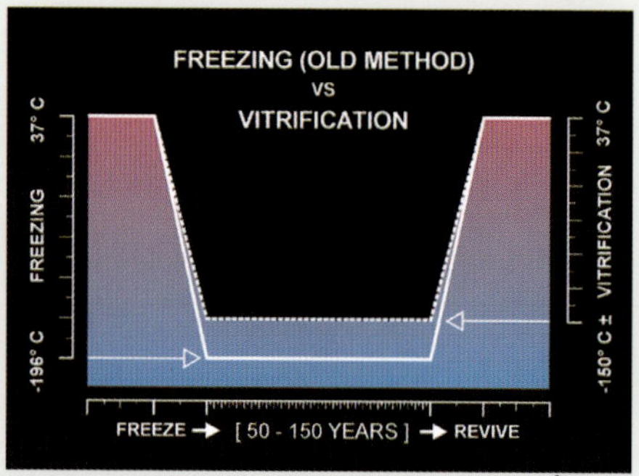

figure [*a*]

BECAUSE OF CRYOPRESERVATION breakthroughs, Timeship is now being built. Previous techniques led to damaging ice crystals, but vitrification prevents the formation of ice. With growing awareness of these breakthroughs, demands for cryopreservation and storage will necessitate the building of new state-of-the-art facilities.

In water, temperature reduction below a certain point results in an abrupt phase transition to a crystal, which we commonly call freezing. In some other systems, this does not happen. Instead, temperature reduction just causes more and more slowing of molecular motion and slowing of chemical reaction rates until a critical temperature is reached. Below this temperature, there is insufficient energy for the most mobile molecules in the fluid to move appreciably. At this temperature, called the "glass transition temperature," the system almost completely loses fluidity and becomes a "solid liquid," which is known as a "glass," and is said to have "vitrified."

Scientists have now developed advanced mixtures of conventional cryoprotectants and "ice blockers" that allow tissues to be vitrified without damage from ice crystals or the vitrifying mixtures themselves.

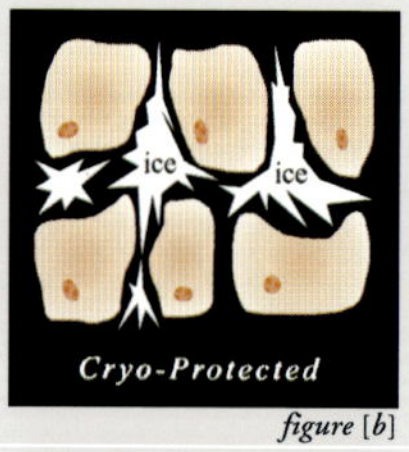

figure [*b*]

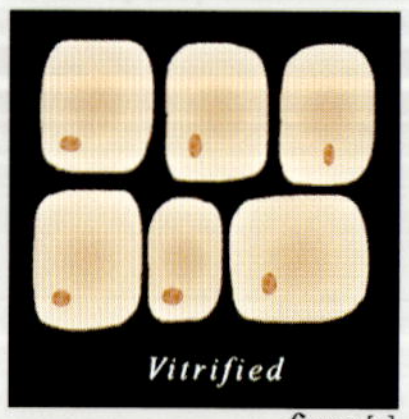

figure [*c*]

[*a*] Temperature timelines for the outdated cryonics process, causing ice damage, and modern vitrification, which prevents ice damage.

[*b*] Cryopreservation with glycerol allows ice damage between cells.

[*c*] Vitrification with no ice crystal damage.

[*d*] Relationship matrix between ice blocker molecular shapes and ice binding ability.

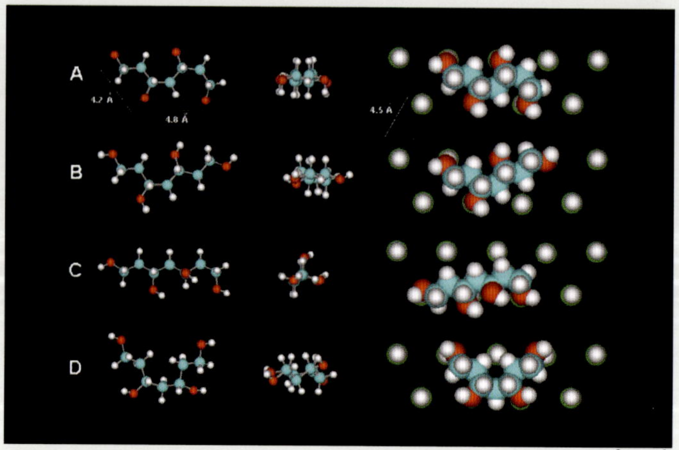

figure [*d*]

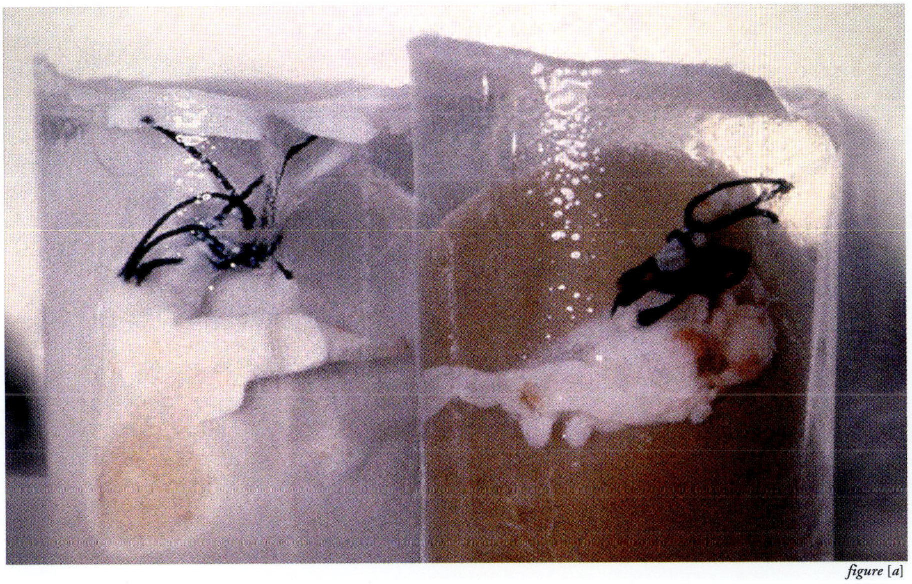

figure [*a*]

[*a*] Left, frozen kidney; right, vitrified kidney. "Thanks to … new technologies, we were able, for the first time, to routinely recover kidneys from a temperature of -45°C (-49°F) and then transplant them successfully," said 21st Century Medicine Chief Scientific Officer, Dr. Gregory Fahy in a recent scientific publication. (The black threads are sutures.)

[*b*] 21CM's Supercool X-1000 is now used in the vitrification process, allowing cryopreservation without ice formation.

TIMESHIP WILL STORE PATIENTS traveling to the future, as well as human cells, tissues, and organs for transplantation. At this time most patients in need of an organ transplant must wait for someone with tissue compatibility to die nearby in order for a fresh organ to be quickly transplanted. Each year, thousands of people die for lack of compatible organs. If organs could be cryopreserved and stored in organ banks, many of those thousands would live. Right now, most organs cannot be successfully frozen due to ice damage. Vitrification will solve that problem.

With its intricate and delicate structure, the kidney is one of the most complex organs, and one of the most difficult to cryopreserve. By contrast, the heart, made of simple muscle, may prove to be one of the easier organs to cryopreserve. Scientists at 21st Century Medicine (21CM) have successfully shown that it is possible to cryopreserve kidneys and they are now working on other transplantable organs.

21CM's Supercool X-1000™ and Z-1000™ ice blockers are new facilitators of vitrification. Even whole organs can now survive vitrifiable concentrations of cryoprotectants. 21CM's new vitrification solutions are finding applications in systems as simple as mammalian ova and as complex as whole mammalian organs.

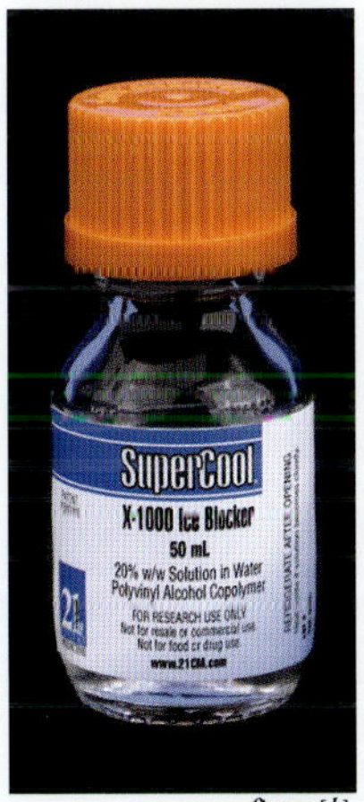

figure [*b*]

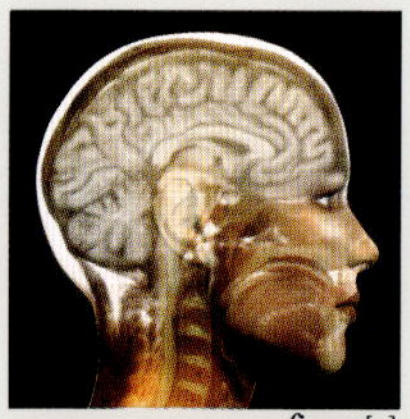

figure [*a*]

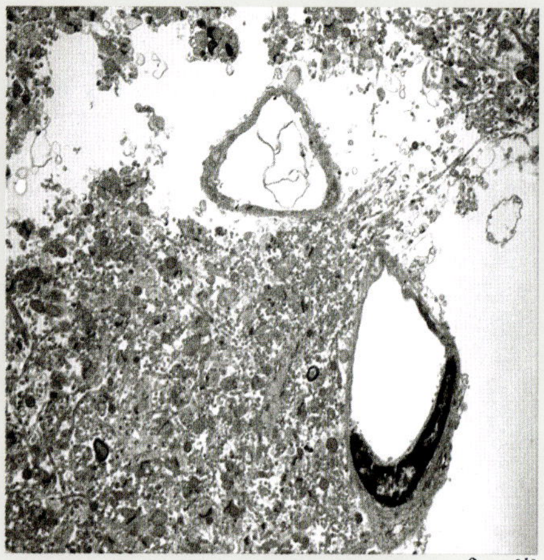

figure [*b*]

figure [*c*]

THE POINT OF HUMAN CRYOPRESERVATION, also known as cryonic suspension, is to preserve one's identity, which might be understood as one's "self" and memories, until scientific advances make possible reanimation in a repaired or new, age-resistant body.

Those who have chosen to be cryopreserved assume that the brain is the seat of this self that they hope will live on after reanimation. Research done to date does in fact indicate that memories reside in the structures of the brain and that these memories remain for some amount of time after the heart stops. This research further suggests that memories will survive the process of cryopreservation and reanimation. For this reason, many patients choose to have only their heads preserved, the head being an ideal protective packaging for the brain. Such patients hope that someday technology will make it possible for new bodies to be regenerated around their brains. A few patients even speculate that in the distant future they will be able to be "downloaded" into some new form of existence that at this time we can only begin to imagine. However, due to newer techniques that lead to less damage to the body during cryopreservation, more patients may now choose to have their whole bodies preserved.

[a] The head serves as an ideal protective container for the stored brain, the seat of the self.

[*b*] Electron micrograph of brain tissue preserved by freezing with limited cryoprotectant, an older method. The white areas show ice crystal damage.

[*c*] Brain tissue preserved by vitrification with no ice damage, only some cell shrinkage that reverses when the vitrification chemicals are removed.

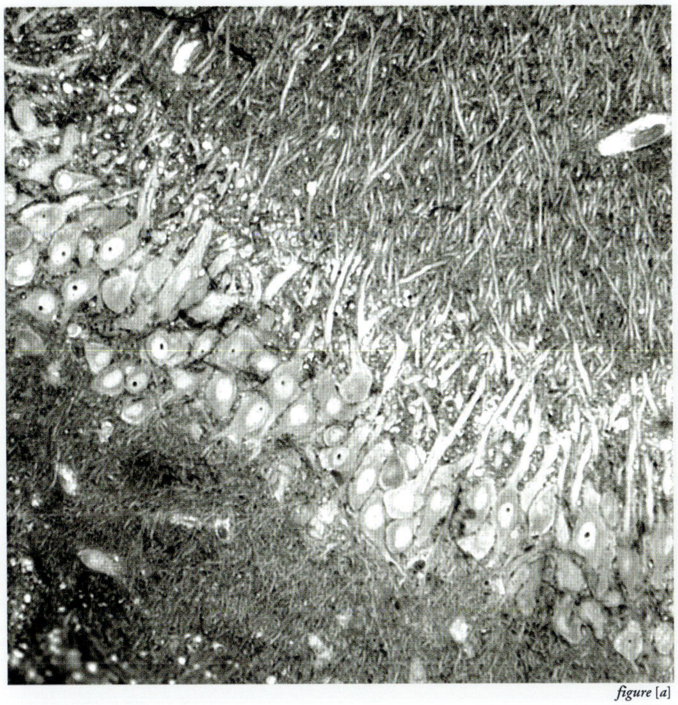

figure [*a*]

[*a*] Superbly preserved structure of mammalian hippocampal slice after vitrification/rewarming. Mean viability scores averaged at almost 100 percent overall.

[*b*] Ice crystals cause damage to tissues and cells with older cryo-preservation techniques. The first patients to be cryopreserved were aware of this problem, but also believed that rapid progress in medical technology would eventually make it possible to repair their damaged cells upon reanimation. Later patients believed that nanotechnology would facilitate this. Now vitrification techniques reduce the need for such repairs.

figure [*b*]

IF THE BRAIN IS the seat of memories and of the self, then it must be cryopreserved in such a way that there is as little damage as possible and it can be successfully reanimated. The brain is therefore a major challenge in cryopreservation. Cryobiologists have long explored preservation at very low temperatures, but have had longstanding difficulties in preserving systems as complex as organs. However, recent studies have proven successful in partially recovering electrical activity and viability in whole brains that have been fully perfused with enough cryoprotectant to vitrify. Similar success after vitrification and rewarming could be verified soon.

A few organs have been able to survive and function when transplanted even after deep freezing and thawing. Therefore, even imperfect vitrification of whole bodies should still preserve much biological integrity in the future.

[*a*] Nano memory storage device. This proposed nanorobot would contain entire libraries and directly interface with neurons, thus enabling people to have constant access to all of the world's information. Of course the data would have to be searchable. Education could then focus on conveying wisdom. Note: the nanotechnology images on this and the following page are artists' conceptions.

[*b*] & [*c*] Proposed cell repair nanorobot control hub, which coordinates work and transmits data to a macrocomputer for analysis. The hub also helps nanorobots navigate within the brain.

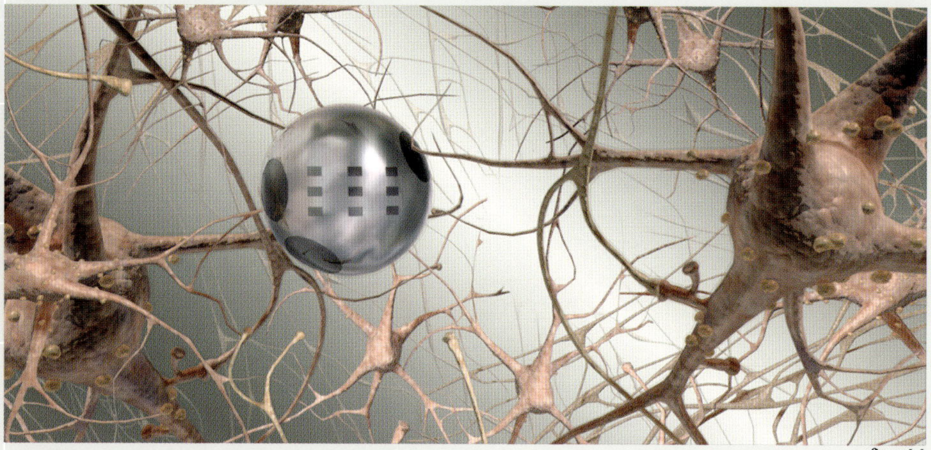

figure [*a*]

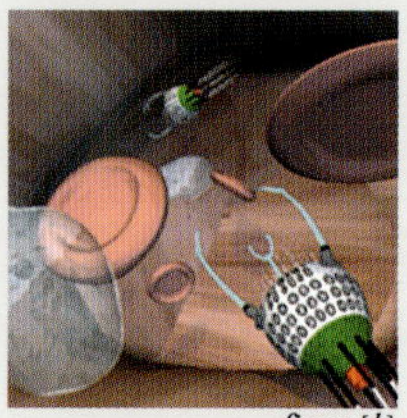

figure [*b*]

FIRST CONCEIVED BY physicist Richard Feynman and elaborated by Eric Drexler in *Engines of Creation,* nanotechnology refers to the manipulation and construction of materials at molecular and atomic scales, and has broad current and future applications. As computer chips get smaller and the capacity of memory media increases, nanotech applications are already important in the computer industry. Future speculative uses include molecular manufacturing—the highly economical manufacture of just about everything in massively parallel nanofactories that will assemble devices, both small and large, one molecule at a time.

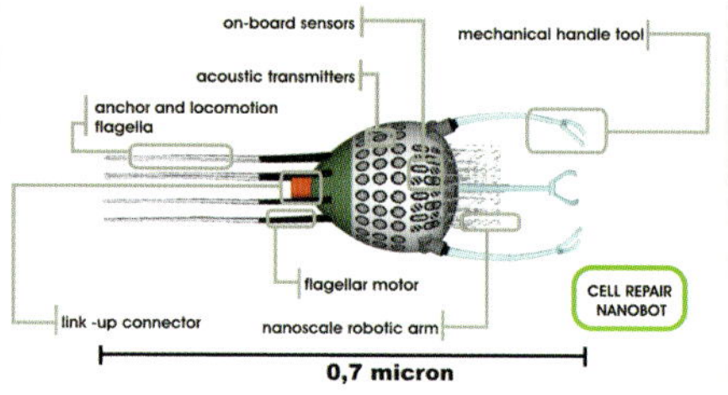

figure [*c*]

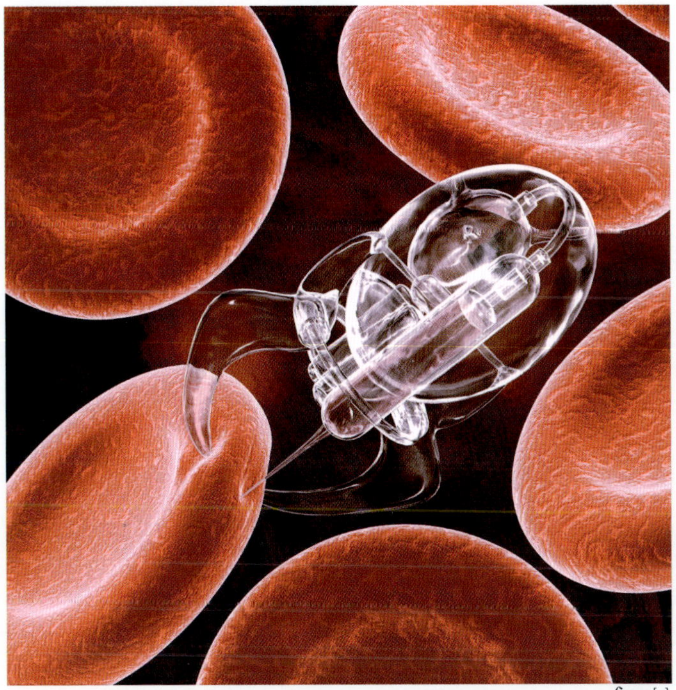

figure [a]

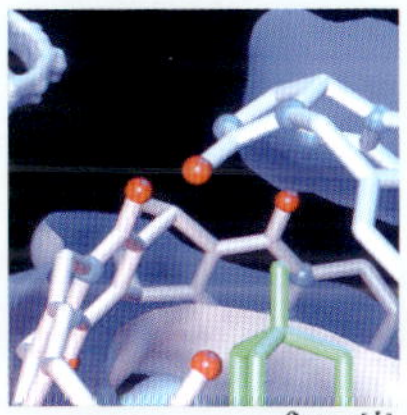

figure [b]

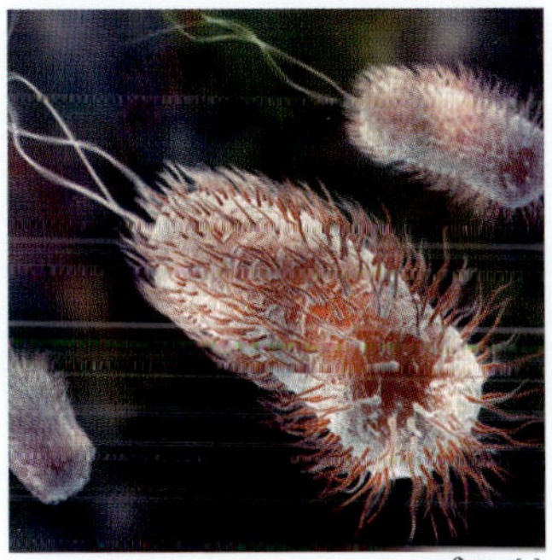

figure [c]

NANOMEDICINE, BASED ON nanotechnology, is of great interest in the field of life extension. Nanorobots with powerful on-board computer controls might be made to enter the body, diagnose diseases, release drugs, repair cells, remove pathogens, and unclog arteries. They may also be able to enhance or even create new brain functions.

When it becomes possible to reanimate many of Timeship's patients, they will have suffered damage both from the illnesses that made it impossible for the physicians of their day to keep them alive, and from the cryopreservation process. Nanomedicine will play a key role in the future "refurbishing" of reanimated Timeship patients. Such nanomedical repairs will be even more important for patients who were cryopreserved before vitrification, and who will need extensive repairs due to damage from ice crystals.

[*a*] Schematic showing proposed nanoinjector roaming through the bloodstream, taking test samples from red blood cells and injecting drugs into them as directed by computer instructions.

[*b*] Components for proteins built by future molecular manufacturing techniques.

[*c*] Proposed nanotech hunter units roaming body fluids to search out and destroy pathogens.

[*a*] DNA sequencing that may eventually lead to gene therapies to stop and reverse degenerative diseases and perhaps aging.

[*b*] Kamato Hongo died in Japan at 116, indicating the limit of natural old age.

[*c*] Bob the Dog spent over 15 minutes without blood circulation and oxygen at normal body temperature before being revived with Critical Care Research's protocol. On revival, Bob's behavior remained the same, including recognizing people. This showed memory can survive "death," and therefore reanimated cryopreserved patients may retain their memories. Bob is alive and well, and he offers hope to heart attack and trauma victims.

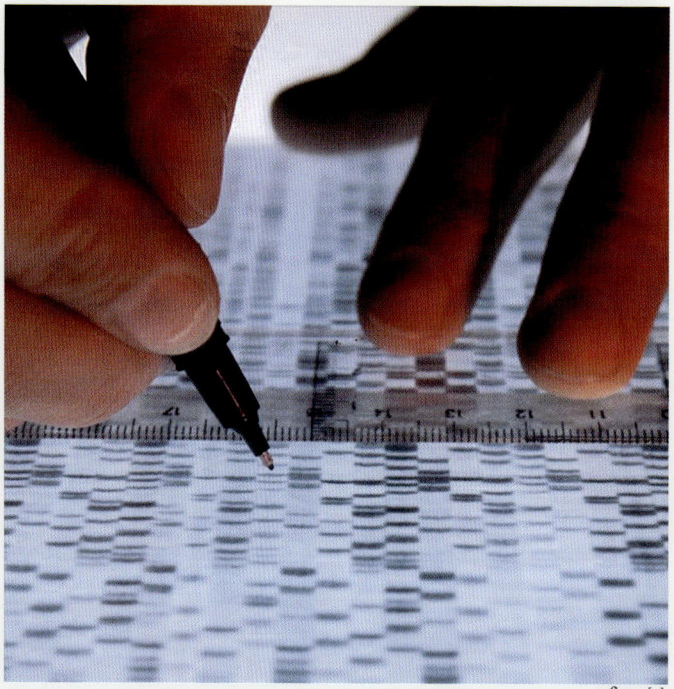

figure [*a*]

figure [*b*]

LIFE EXTENSION RESEARCH is intended not only to extend one's life span, but to assure that it is spent in youthful vigor. Two companies doing this research are BioMarker Pharmaceuticals and Critical Care Research. BioMarker uses DNA microarrays (gene chips) to identify "biomarkers" of aging through the analysis of thousands of genes at a time. We know caloric restriction extends life span in mammals, so BioMarker is seeking therapies that produce the same effect without perpetual semi-starvation. The results could also prevent diabetes, cancer, cardiovascular disorders, and neurodegenerative diseases. Critical Care Research has developed drugs and a hypothermia system for use in trauma units, ambulances, and military field hospitals after a heart attack or trauma. The protocols could triple the "non-treatment time" before brain damage occurs. These protocols will also be used in cryopreservation.

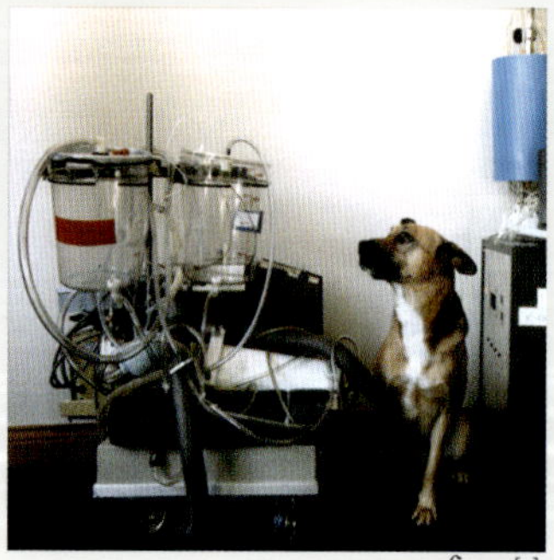

figure [*c*]

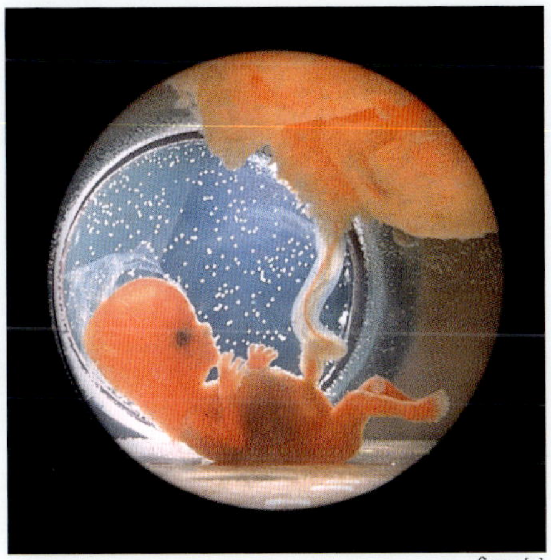

figure [*a*]

[*a*] A fetus showing life in formation, and evoking the Star Child, the new being at the end of *2001: A Space Odyssey.*

[*b*] A yellow seahorse whose DNA will be cryopreserved with that of other endangered species on an "ark" in the U.K., similar to the one at Timeship.

[*c*] Many people have only their brains, the seat of the self and of memory, cryopreserved.

figure [*b*]

THE THREE LEADING TECHNOLOGIES of our time are electronics, nanotechnology, and biotechnology. Timeship is at the heart of biotechnology. Its mission is research into life extension and preservation of biological materials, including DNA from extinct and near extinct species, organs for transplantation, and patients traveling to the future for reanimation—and, hopefully, practical immortality (death only by trauma or choice). People in the future may wonder how we ever put up with aging and death.

These technologies have the potential in the next few decades to bring about greater changes than those of the past few centuries. Many technologies are developing at an exponential rate, doubling every few years or even months, and we may be approaching a "Singularity," a term indicating a time when change happens so fast we are not able to comprehend it from current perspectives.

figure [*c*]

03 ARCHITECTURE

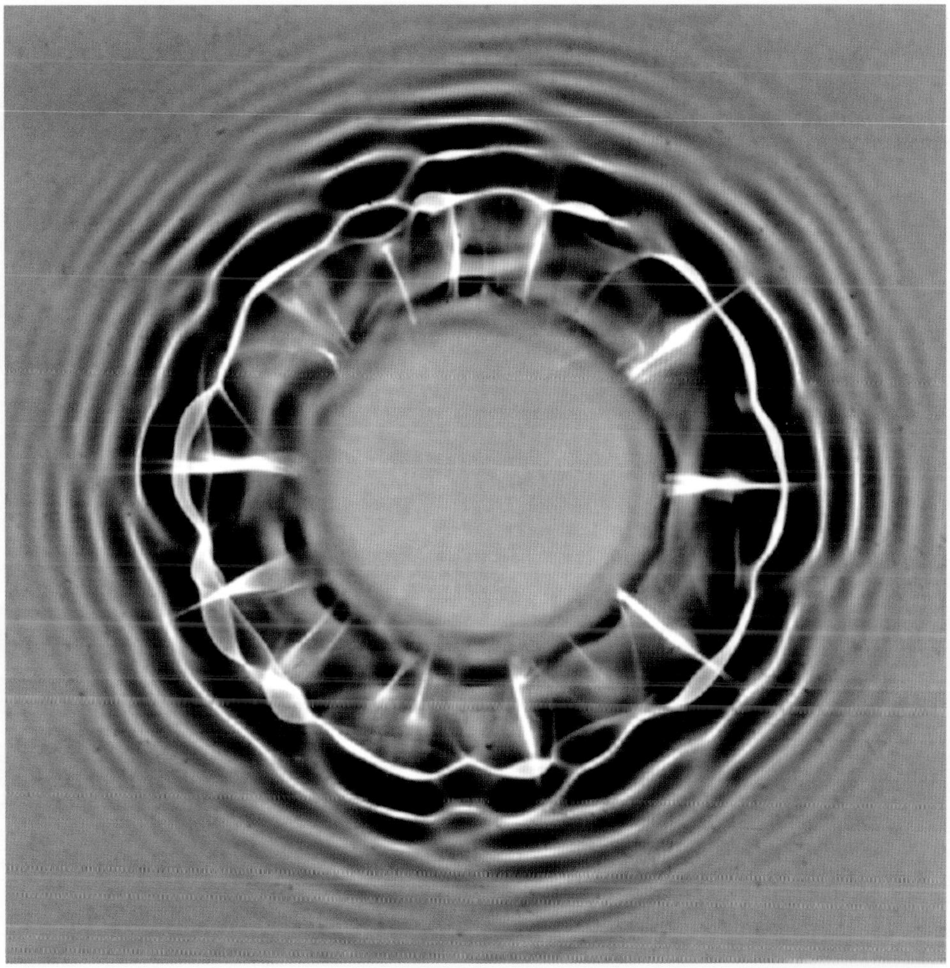

ARCHITECTURE

A "BUILDING" HOUSES THE FUNCTIONS for which it is designed, but "Architecture" must also help us understand and absorb the meanings of those functions. Timeship's functions—radical life extension and cryogenic travel to the future—challenge our traditional ideas of human life as typically limited to seventy or eighty years. In fact, its quest for practical immortality may completely change our understanding of what is "human." In order to fulfill Timeship's role of helping us deal with these challenges, architect Stephen Valentine has incorporated in the design symbols of the human place in the world and of regeneration drawn from many cultures and from the breadth of history. Among these symbols are sacred geometry, biological patterns, planetary orbits, the mandala, water, architectural monuments, and the Phoenix.

Symbols are our most primordial means of expression, connecting us to that which transcends logical description. Unlike signs, which are arbitrarily assigned and are learned, symbols are universal and they inherently resonate with all human beings. They convey an understanding of reality and a way of looking at and making sense of the world that cannot be directly expressed in words. There is a human need to find patterns in experience; symbols address that need, helping us chart our evolution as a species and our relationships as individuals to our culture, to other cultures both present and past, and to the cosmos. Symbols constitute a universal language and their particular application to architecture was studied extensively by the late architect and professor, Mimi Lobell.

Primary among the symbols chosen by Valentine are those we group together under the term "sacred geometry." Because of its universality, sacred geometry has the power to tie together humans, nature, and the universe. Examples of sacred geometry include the circle; the square; the squared circle; the Fibonacci Series; gnomonic expansion; the spiral; and the Golden Ratio, Section, and Mean.

The significance of the square includes firmness, stability, and permanence. When used in sacred architecture it signifies the worldly. The circle represents totality, wholeness, and completion, as well as the self-contained, the unmanifest, the infinite, and eternity. When used in architecture it signifies the otherworldly. The dynamism of the circle suggests endless movement and recurrence. Since it has no beginning and no end, the circle also represents going beyond time and space. Squaring the circle refers to geometric operations that attempt to construct a square with a perimeter (or area) equal to the circumference (or area) of a particular circle. This can be done only approximately because the circumference of a circle relies on pi (π), an irrational number, and therefore the basis for the circle being associated with mystery and the realm of the sacred.

By using the formulas of sacred geometry, squaring the circle creates the Four Corners of Timeship, which symbolically represent the four traditional elements: water, fire, earth, and air. These Four Corners are used as mechanical rooms, vertical circulation, and security lookout towers. These four elements can even be seen today in American pop culture in the Marvel Comics super heroes, The Fantastic Four: Stretch, Torch, Thing, and Invisible.

The Golden Ratio is based on an irrational number, approximately 1.618, that possesses interesting mathematical properties. It generates shapes that are considered aesthetically pleasing, combining harmoniously with other forms and reflecting the balance between symmetry and asymmetry in nature.

Sacred geometry refers to forms, patterns, intervals, and relationships that are found in many cultures and exist in our bodies, in our perceptions, in living creatures, in human creations, and in the structures of the universe. It is contained within the yin and yang, the cross, and the mandala, and is seen in religious and spiritual traditions throughout history. Modern physicists are even beginning to see the universe more in terms of patterns of mind than matter and energy. The quantum physicist James Jeans famously stated that "the universe begins to look more like a great thought than a great machine." The notion of the universe being patterns of mind is similar to the position of many ancient cultures that examined reality through the metaphors of geometry.

Also important for helping us position ourselves in the cosmos is the mandala. The term mandala is Sanskrit for circle, but it is commonly applied to a geometric pattern that represents the structure and order of the cosmos. At the same time the mandala is a model of the human mind and therefore a meditation tool for putting the self in harmony with the larger order of things. The relationship of the parts of the mandala to the whole is symbolic of the desired relationships of the parts of the mind to the whole.

Mandalas are typically symmetrical and concentric, and are often used as plans of palaces, thus making them particularly appropriate as symbolic forms related to Timeship. As the plan of Timeship reveals, the geometry of the perimeter wall, in combination with the core of the building, form a mandala with Aristotle Plaza and the Phoenix Cone at its center. This is a powerful metaphor, symbolic of a faith in humanity and rational thought, and thus speaks of a belief in ourselves. This moves beyond religion and puts responsibility in our own hands. Aristotle Plaza, with its surrounding laboratories, represents the point of stillness about which pivot various degrees of creation. The perimeter wall provides physical and metaphysical security, as symbolized by the mandala, which protects the area it circumscribes against external forces. The analytical psychologist Carl Jung saw the mandala as representative of the wholeness of the self, and often had his patients paint mandalas to externalize the degree of fragmentation or integration they were experiencing.

Water is another symbolic source for Timeship. Its reference to immortality and limitlessness evokes the underlying theme of Timeship: the regeneration of the human species and the belief in the endless potential of human knowledge. This can be visualized by the often seen high speed photographs of expanding water ripples caused by a water droplet falling into a tranquil pond, and the ensuing column of water projecting from the surface in proclamation of its origin or birth. The water droplet represents the moment that life emerges. It is measured in nano time, while the water ripples continuing on to infinity are measured in macro time. This image has served as an inspiration for the design of Timeship, as can be seen in the plan.

Besides geometry and nature, Timeship also draws on the richness of our architectural heritage, including the Benben stone, the predecessor of the Egyptian pyramids and the inspiration for the Phoenix, the symbol of rebirth. Other historical architecture on which Timeship draws includes Stonehenge, the Greek temple, the Islamic mosque, the Gothic cathedral, and Indian observatories.

The great modern architect, Mies van der Rohe, wrote: "Architecture is the real battleground of the spirit. Architecture depends on its time. It is the crystallization of its inner structure, the slow unfolding of its form." What is the spirit of our time? We are in the midst of the most significant period of technological and therefore cultural change in history. Developments in astrophysics, communications, computing, and nanotechnology are radically altering the way we understand and control our world, even down to the level of individual subatomic particles. But far more importantly, biotechnology is changing our understanding of and ability to control life itself. Through its role in extreme life extension and cryopreservation for travel to the future, Timeship is the first work of architecture to be fully embedded in and a vehicle for these changes. Through its symbolic incorporations, it will help us absorb the meanings as well as the implications of these changes.

[*a*] Salvador Dalí's famous 1931 painting, *The Persistence of Memory,* showing melting clocks and indicating that time is not the fixed mechanical thing we used to imagine. What is the role of memory in our experience of time?

[*b*] Annual tree rings on a cross section of a log. Is time linear? Is it cumulative? Can it be stopped? What are the implications of practical immortality and how would it change our notions of time and the course of our lives?

[*c*] Stop-action photo of a diver. Is time a continuous flow or is it made up of discrete moments?

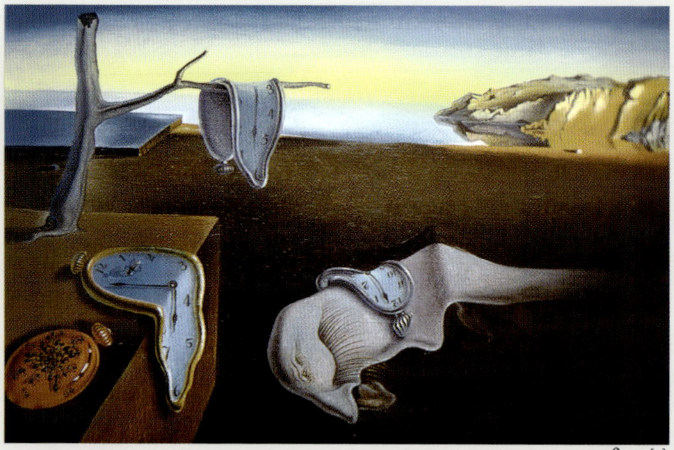

figure [*a*]

figure [*b*]

TIMESHIP WILL EVENTUALLY make the distant future accessible to its patients just as some form of suspension will make distant planets and stars accessible to astronauts. Our notions of time will have to change completely. Time is one of nature's mysteries. Psychological time is variable, but aging is relentless. However, what will happen to our sense of time if aging is stopped—a goal of Timeship's sponsors—and we achieve practical immortality?

figure [*c*]

figure [*a*]

figure [*b*]

[*a*] The Egyptian Pyramids, our most enduring symbol of timelessness, carried their passengers to another realm.

[*b*] The greatest measure of time, Stonehenge, is both a calendar and a calculator, able to predict solar and lunar eclipses as well as the changing seasons.

[*c*] Heavenly objects are sighted along the edges of these elegant structures of an open-air observatory that was built in 18th-century Jaipur, India by the Maharaja Jai Singh II.

TIMESHIP ENGAGES the rich themes of both time and timelessness as do many of the most recognized architectural monuments of the past. The Timeship building will be operative for possibly hundreds of years and a number of its passengers will be maintained there for as long. Because of this, Stephen Valentine has chosen a timeless approach to the design and the symbolic themes, and has made reference to timeless buildings.

Timeship is a vehicle to carry its passengers to the future, making it a time machine. But Timeship is also of its time, housing the most advanced biological technologies representative of our age.

figure [*c*]

figure [a]

[*a*] Time lapse photo of the night sky over an observatory showing stars "circling" the North Star.

[*b*] The rings around Timeship can be projected outward indefinitely to evoke the famous Charles and Ray Eames film, *Powers of Ten,* which shows our location in the universe by taking ten tenfold jumps out beyond our galaxy, and down into subatomic particles.

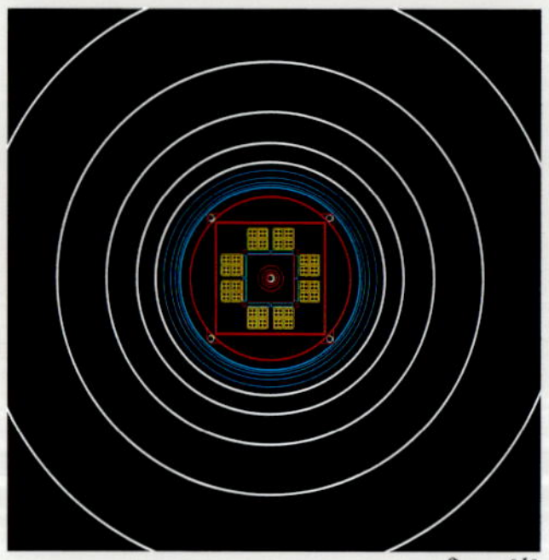

figure [b]

A MAJOR ROLE OF ARCHITECTURE throughout history is to frame the human place in the universe. Timeship is positioned on a World Axis, an archetypal symbol first appearing in ancient Shamanism and continuing throughout the history of architecture and the arts. We see it in the pyramids of Egyptian and other cultures: the Obelisk in the center of Bernini's St. Peter's colonnade, the Kaba in Mecca, the Washington Monument, Brancusi's sculpture the *Endless Column*, and the modern skyscraper. The World Axis marks the center of the world, around which all else rotates, and ties together our realm with the upper and the lower realms. But, of course, any place can be the axis of the world.

The axis of Timeship is derived from the four cardinal points. These points are utilized in establishing its layout, thus orienting the building in relation to its site, the region, the Earth, and the universe.

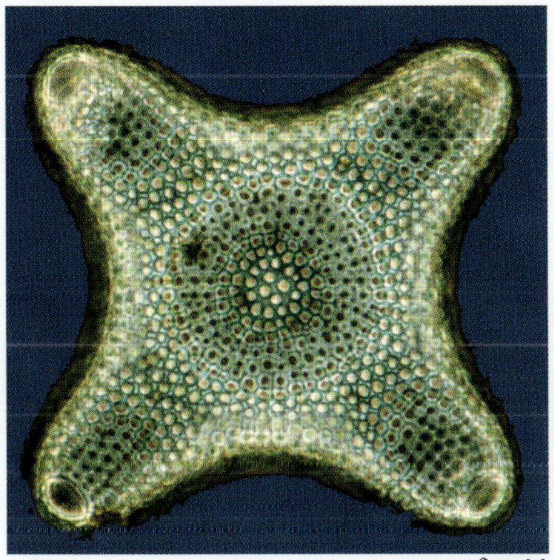

figure [*a*]

figure [*b*]

[*a*] & [*b*] Diatoms with square and pentagonal symmetry. The elegant symmetries of diatoms are a result of the patterns in which silicon atoms can be joined. Valentine studied diatoms and other symmetries found in nature and the cosmos as an inspiration for the symmetries of the Timeship building.

[*c*] Plan of Timeship reflecting symmetries found in nature.

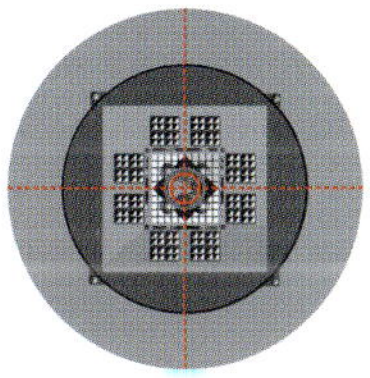

figure [*c*]

IN THE NINETEENTH CENTURY, architects began to turn away from history as a source for their designs, looking instead at nature. For example, the French architect, Viollet-le-Duc, studied the shapes of crystals, saw them repeated in mountain formations, and used them to inspire his architecture. He also pioneered in the use of iron as a structural material, proposing designs based on the forces moving through a building.

The search for inspirations from nature extended to the microscopic realm with the study of diatoms and radiolarians. The beauty, symmetry, intricate patterns, and endless variety of these creatures have fascinated naturalists since the invention of the microscope, and more recently have inspired morphologists, engineers, and architects. The poetic engineer, Robert Le Ricolais, longtime colleague of architect Louis Kahn, taught generations of students to study the elegant, efficient structures of radiolarians.

Many architects have also been inspired by biomathematician D'Arcy Thompson, whose book, *On Growth and Form*, shows the roles of physics and mechanics in determining the form and structure of organisms. Principles of nature, along with symbolism, were a major influence on Valentine in his design of Timeship.

figure [*a*]

[*a*] High-speed photography in nano time and time-lapse photography in macro time have opened up new views of our world. Here, a crown forms from a drop of milk.

[*b*] The outer rings of Timeship provide physical defenses while at the same time symbolically represent a moat and the rings moving outward from a drop striking a body of water.

[*c*] A water drop striking a smooth water surface showing the column, the secondary drop generated off its tip, and the waves generated out from it. This image is a direct inspiration for the form of Timeship.

WATER, OUR COMMON DENOMINATOR, is a giver of life, symbolizing the interdependence of all things. Rain is associated with fertility, dew with benediction and the light of dawn, and running water with life. Immersion in water is associated with purification and spiritual rebirth. Water is the beginning and end of all things, and has also been identified with intuitive knowledge and the unconscious.

Water is basic to Timeship with its reference to immortality and limitlessness, as can be seen in the expanding ripples caused by a water drop falling into a tranquil pond and the ensuing column projecting from the surface, which is mirrored in the plan of Timeship.

figure [*b*]

figure [*c*]

figure [*a*]

[*a*] At the center of Aristotle Plaza is the Phoenix Cone, which represents a flower, the Benben stone, and also a column of water. Mirrors at the perimeter are angled to reflect the cosmos (clouds, sky, stars, and planets) into the plaza. The four walls represent Aristotle's Four Causes. Surrounding Aristotle Plaza and covering the patient storage areas are the research laboratories illuminated by prismatic skylights. The laboratories at Timeship will conduct biotech, biomedical, cryonic, life extension, and related research.

[*b*] Timeship and its surrounding rings.

figure [*b*]

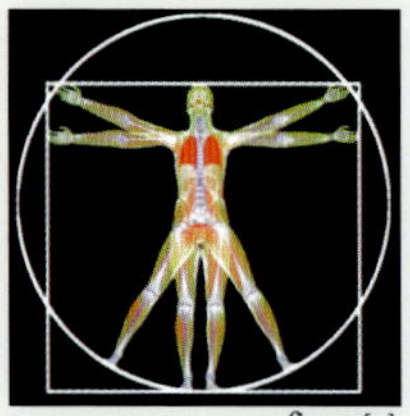

figure [a]

THE CONCENTRIC CONFIGURATION of Timeship with the patient storage area surrounding the center of the building, and in turn surrounded by protective walls, lends itself to concentric sets of squares and circles. This concentric configuration is not only reflective of the squared circle, but also of the mandala. In addition it serves the functional role of contributing to the highly secure environment of the patients and biological materials cryopreserved within Timeship.

[*a*] Da Vinci's Vitruvian Man as MRI, showing harmony between man and geometry.

[*b*] This geometry of the squared circle contains the pentagon, which represents life through a fivefold symmetry found in living organisms.

[*c*] The permanence, stability, and enclosure of the square are played against the wholeness, completion, and infinity of the circle.

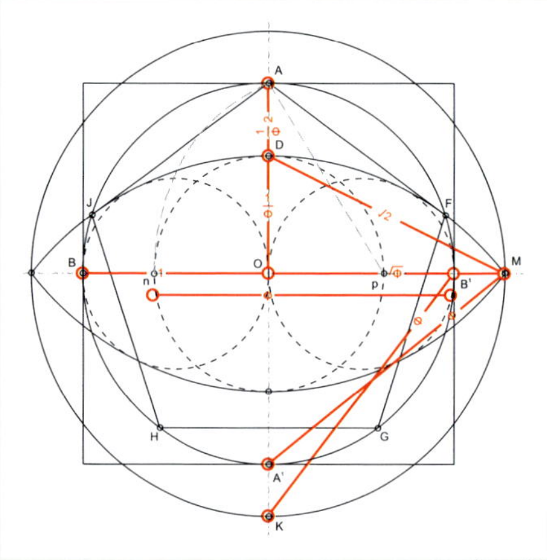

figure [b]

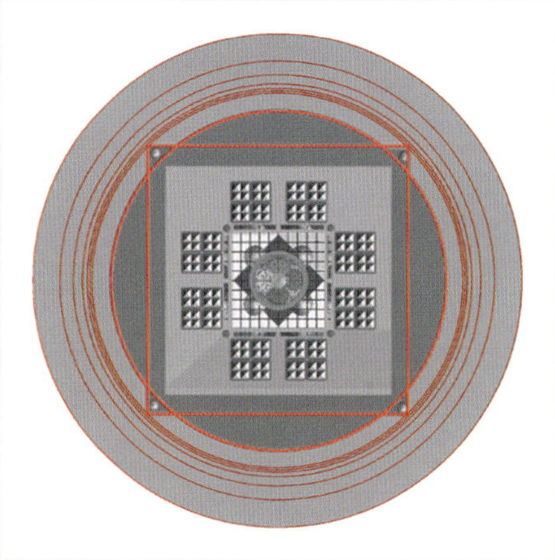

figure [c]

SQUARING THE CIRCLE is creating a square with the same area (or perimeter) as the area (or circumference) of a given circle. With calculus we can now solve this problem, but for centuries it evoked mystery and our connection to the greater cosmos.

The Roman architect Vitruvius advocated proportions in architecture based on the human body. Leonardo da Vinci followed Vitruvius's descriptions to create his "Vitruvian Man" set in a squared circle.

figure [*a*]

figure [*b*]

THE HARMONIC PROPORTIONS of the Fibonacci Series underlie Timeship's design. The series is recursive—one starts with two numbers and produces the next by adding the two previous numbers. For example, 0, 1, 1, 2, 3, 5, 8, etc. The series can also be the basis of a spiral. Proportions are based on ratios, which represent our ability to differentiate through our senses, which perceive through differences. This is a basis of conceptual judgment. Proportions are more complex than ratios, as they involve finding relationships. The proportions of Timeship give users an unconscious sense of order and connectedness.

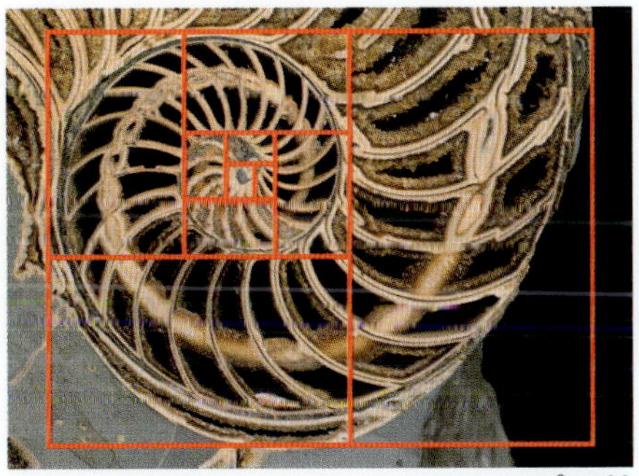

figure [*c*]

[*a*] Spira mirabilis from the ceiling of a mosque in the Taj Mahal complex.

[*b*] Cactus showing a Fibonacci Series spiral pattern that grows out of the Golden Proportion. The expansion of the universe, in which fundamental Unity is found in new creations, is expressed in The Golden Proportion.

[*c*] Little changed for over 100 million years and the subject of poetry ("This is the ship of pearl, which, poets feign, Sails the unshadowed main…"), the shell of the chambered nautilus, grows as the creature moves to new quarters. Its proportions bear a mathematical relationship to the logarithmic spiral that inspired the volute of the Greek Ionic order as well as the proportions of Timeship.

figure [*a*]

[*a*] The swirls in this figure aid meditation on the bindu, the still point at the center.

[*b*] The center of the Sri Yantra comprises intersecting upward and downward triangles, symbolizing the union of male and female and forming a fractal-like pattern.

[*c*] Ancient Mexican Aztec square mandala representing the universe centered around a god of time.

WHILE THE MANDALA is most often associated with Eastern cultures, it has appeared in most cultures throughout history. Typically it is a series of concentric circles and squares surrounding a central symbol that can be figurative and usually refers to unity. Its center represents a safe haven and a sacred space protected from exterior forces. As a model of the mind as well as the cosmos, the mandala is used in some forms of meditation by many cultures to achieve a sense of unity, wholeness, and perfection. Even patients of modern psychology have used this symbol as a way to make sense of a fragmented world.

figure [*b*]

figure [*c*]

THE MANDALA PLAYS a vital role in the design of Timeship. The geometry of the perimeter wall in combination with the core of the building form a mandala with Aristotle Plaza as its center. What in the past was the abode of the gods is fittingly today the domain of the scientists and researchers who make life extension possible. This is symbolic of faith in humanity and rational thought, and thus speaks of a deep belief in ourselves.

figure [*a*]

figure [*b*]

figure [*c*]

THE MANDALIC NATURE of the Timeship plan begins with the outer rings of bermed earth. Next is the circular outer wall that is "squared" by four security towers. Inside the wall is infill, which protects the square patient cryostorage area topped by the research laboratories. At the center is Aristotle Plaza. The plan of Timeship reflects both the symbolic mandala and the need for a high degree of security for its cargo traveling to the future.

[*a*] A 14th-century Christian mandala with a cross and angels turning the wheel of the universe, similar to the Buddhist Wheel of Life.

[*b*] & [*c*] Although it was not intentional on the part of the architect, there is a remarkable similarity between this mandala on a Chinese mirror and the plan of Timeship. The plans of temples are often mandalas, with a center and gates on the cardinal sides. Although Timeship avoids gates in its walls in favor of a secure underground entrance, it retains the cardinality. Timeship's plan shares with these mandalas an approach in which successive rings, representing various entities, continually radiate out from a generative center.

[*a*] The Mosque of Ibn Tulun in Cairo is an island of quiet in a hectic world. Its dome represents the meeting of heaven and earth, mirroring Timeship's Phoenix Cone.

[*b*] Chartres Cathedral's rose window is a mandala, generated from a spiral of the Fibonacci series, with spiritual, meditative, and emotional significance. It allows luminous transcendence to penetrate the building.

[*c*] The Mihrab Ceiling of the Cordoba Mosque in Spain with a star pattern generated by eight intersecting arches that join square, star, octagonal, and round forms. Its centrality mirrors that of Timeship.

figure [*a*]

figure [*b*]

CHRISTIANITY AND ISLAM are both Biblical religions. Islam's origin in the hot, dry climates of the Middle East resulted in an architecture of walled enclosures. Islam's anti-iconographic tradition led to an embrace of abstract geometric patterning in the arts. There is also an emphasis on two axes in its mosques: the horizontal pointing to Mecca and the vertical pointing to Allah.

The great Christian cathedrals also pointed skyward. They were dedicated to Mary, mother of a dying and resurrecting deity. Both of these traditions inspired the design elements of Timeship, which has strong axes and intense geometry, and is dedicated to reanimation.

figure [*c*]

figure [a]

[a] Research at Timeship will change everything. This time there is no oracle. We ourselves must understand and master these new technologies, as well as confront their ethical implications, debate their meanings in the greater community of ideas, apply them, and take charge of our own destiny.

[b] Temple Pronaia Athena. Some Greek temples were circular, reflecting the dome of the sky. It is believed the Delphic Oracle, who was visited by many great figures of ancient Greece, breathed a hallucinogenic gas from geological faults under the site to see the future.

THE ANCIENT GREEKS, when faced with such challenges as wars or the founding of new colonies, consulted the Oracle at Delphi. Delphi was revered throughout the Greek world as the omphalos, the center of the universe, and the oracle was a priestess of Apollo.

Today, we are faced with the challenges of responding to new technologies, including nanotechnology, electronics, computing, biotech, genetic engineering, and medicine. They have the potential not only to bring us great benefits, but perhaps even to change our understanding of what it means to be human. Timeship will be a major center of these changes. Now we must be our own oracle.

figure [b]

figure [*a*]

[*a*] The Phoenix, which appears in many cultures.

[*b*] A reconstruction of the temple of the legendary Benben stone, later the capstone of the pyramids.

[*c*] Computer model of the center portion of Timeship showing the Phoenix Cone comprised of two intersecting cones. One is upright in gold-mirrored glass and symbolizes the sun and the rising Phoenix. A smaller inverted tensegrity cone intersects the larger cone and symbolizes a meteorite and the descending Phoenix. It is clear, allowing a view from the atrium up through the oculus to the heavens.

IN ANCIENT EGYPTIAN MYTHOLOGY, the Bennu Bird visited every 500 years, trailing a fire across the sky. Later, it was symbolically represented as a conical-shaped piece of metal, and thought to have been the prototype for the Benben stone. Obelisks and capstones of the Great Pyramids were based on its design. The Greeks named it the Phoenix, a symbol of immortality that represented the triumph of eternal life over death. Other cultures, including those of China and India, had similar legends. Timeship's central cone represents both the Phoenix and the sun. The sun symbolizes life, energy, and the center of the universe.

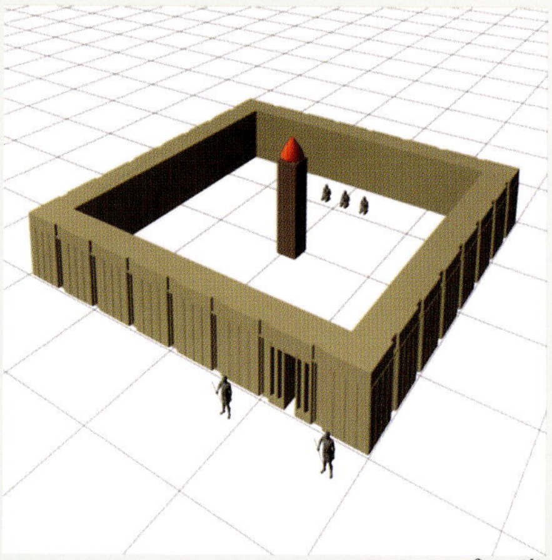

figure [*b*]

figure [*c*]

figure [a]

[*a*] The Hubble Space Telescope and planetary flybys are making our spectacular solar system more visible. One of its most striking features is Saturn's rings, which are representative of concentric patterns found throughout the cosmos.

[*b*] Our understanding of the planetary orbits, which inspired the rings around Timeship, originates with the work of the astronomer and mathematician Johannes Kepler.

[*c*] Aerial view of Timeship showing the rich geometries of its plan, including the squared circle and the rings positioned in the Fibonacci Series.

figure [b]

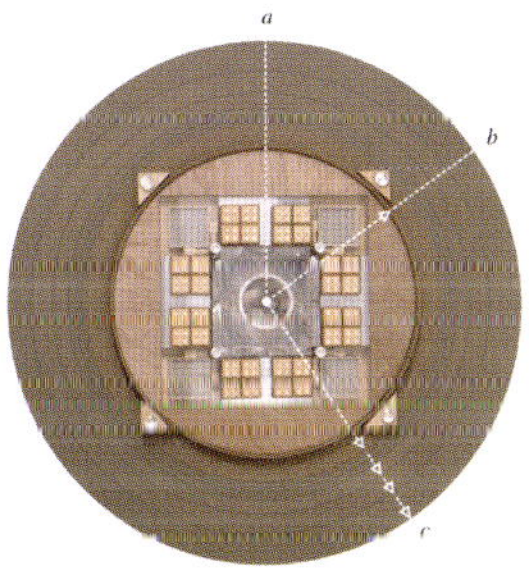

figure [c]

FIRST OBSERVED BY GALILEO in 1610, Saturn's rings are among the most beautiful phenomena in our solar system, and also among the most mysterious, as we still do not fully understand them. They appear to be of relatively recent origin, are influenced by gravitational and tidal forces, and are actually composed of multiple smaller rings with gaps between them. The spacing of the rings and gaps, like planetary orbits, as well as the orbits of moons, reveal proportions that are found throughout nature and that are used in Timeship. Along with the planetary orbits, Saturn's rings are an inspiration for the rings of steps around Timeship that not only provide security by keeping threats at a distance, but also symbolize a moat, water ripples, and ultimately the order of the universe. The rings plus the outer wall represent the orbits of the planets, while the Phoenix Cone represents the sun.

[*a*] The exterior wall of Timeship has a security role and is covered with a pattern evoking flames and woven baskets. The act of passing through fire symbolizes moving to another state, pertaining to the future realm Timeship's reanimated patients will enter.

[*b*] The outer four corner towers of the perimeter wall are for service and security, and also represent the four primal elements: earth, air, fire, and water. Sunlight is directed from these four towers to an opening in the Phoenix Cone where a series of prisms then direct light down into the center atrium.

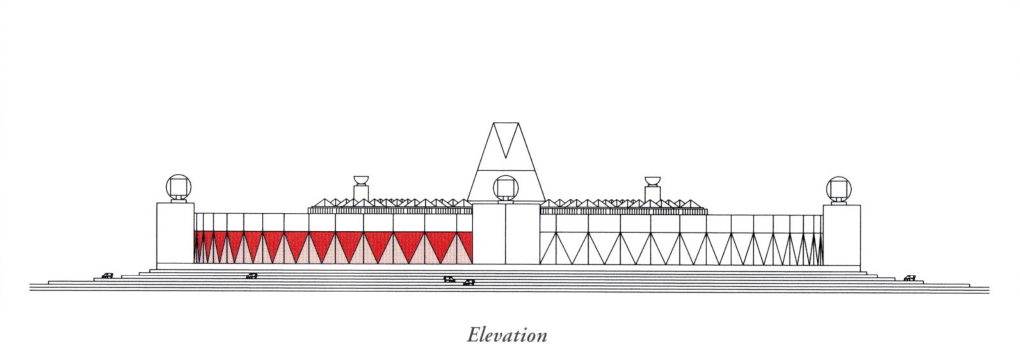

figure [*a*]

figure [*b*]

THE GATELESS OUTER WALL of Timeship is made of fortified concrete and sheathed in stone with a zigzag pattern, an ancient symbol for fire and transcending the human condition. Crossing this perimeter wall through a highly secure underground passage into the interior of Timeship serves as an initiation representing hope for new life.

The symbolism of the squaring of the circle is embedded in the geometry of the perimeter wall. The perimeter wall represents the infinite, the unmanifest, and the immeasurable. The wall also evokes the image of a woven basket, a symbol of rebirth and immortality in many cultures.

figure [*a*]

[*a*] The concentric organization of a flower evocative of the center of Timeship, although the flower's organic irregularity contrasts with Timeship's precision and high-tech materials.

[*b*] The Phoenix Cone at the center of Timeship can be thought of as representing a flower's pistil, but also the sun, the source of all life. The sun's daily rebirth is one source of the legend of the dying Phoenix that rises from its ashes. In addition, the Phoenix Cone represents a column of water resulting from a drop striking a calm surface.

[*c*] A poppy flower's stamens and pistil.

ARISTOTLE PLAZA has the form of a flower, which is androgynous. The flower is the reproductive organ of many plants and produces seeds—the next generation—to spread across the landscape. Each flower consists of a floral radial axis upon which are borne the essential organs of reproduction—stamens and pistil: male and female parts. The Phoenix Cone at Timeship represents the pistil, and the obelisks at each corner represent the stamens.

The reproductive metaphor of the flower is appropriate to Timeship, which is not only a vehicle for the rebirth of its patients, but also will be a major biotech research center that may lead to the birth of the next phases of humanity or even to a post-human stage as biological technologies, including the reengineering of DNA and the joining of life and electronics, make possible new forms of existence that we today cannot even imagine.

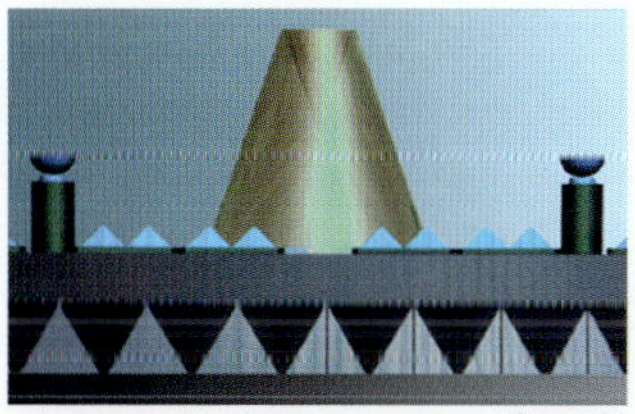

figure [*b*]

figure [*c*]

figure [*a*]

figure [*b*]

AT THE CORE OF TIMESHIP is Prometheus Plaza, which houses the energy source for the project and can be thought of as the "engines of creation." The name of the plaza derives from the Greek mythological story of Prometheus, who brought fire to the earth after stealing it from the gods, and represents the coming of age of humanity as it gained the power to take charge of its own existence.

This plaza is one of the magical spaces in the building, suggesting the wonders of our world and the things in life that give us a sense of awe. It is a reflective place that addresses what makes life worth living, and the importance of the choice of life extension. The floor is transparent so that one is aware of the energy of the symbolic fire below, a reference to the primordial fire and its associations with both the beginning of life and regeneration. An opening above allows one to look up and see the stars.

figure [*c*]

[*a*] The Pantheon in Rome lit solely by the oculus at the center of the dome.

[*b*] Looking up from Prometheus Plaza through the oculus of the Phoenix Cone of Aristotle Plaza into the sky.

[*c*] Looking down into Prometheus Plaza. The spheres are LN_2 storage tanks. The floor is transparent so one sees the energy of the "fire" below as the source of life and regeneration.

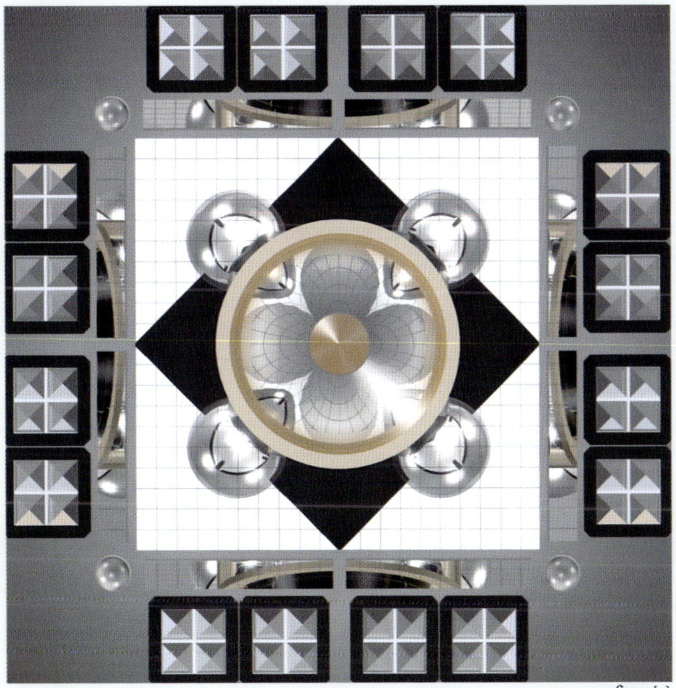

figure [*a*]

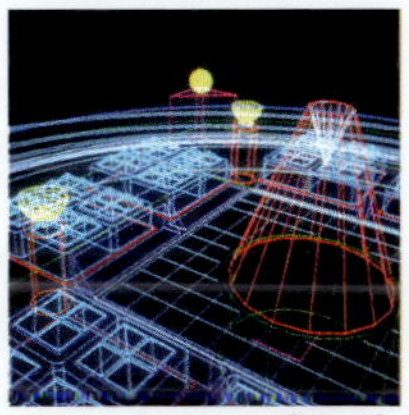

figure [*b*]

[*a*] Looking down at Aristotle Plaza. The form is a mandala reflecting Aristotle's Four Causes—Material: that "out of which;" Substantial: that "by which;" Efficient: that "into which;" and Final: that "for the sake of which," something is made.

[*b*] Computer wire frame design study of the plaza with the Phoenix Cone in red.

ARISTOTLE PLAZA is Timeship's central gathering place as well as a place of rest and contemplation for the research scientists. The referants for the Phoenix Cone at the center include the Phoenix and the sun, and it is separated from the glass plaza that it penetrates from below. Aristotle Plaza is filled with a swirling cloud created by a cloud-generating system. This provides cooling and also symbolizes an abode of the gods, now occupied by the scientists bringing us practical immortality. The plaza has the form of a geometricized flower, and lanterns symbolizing wisdom and intellect top the obelisks at the four corners. On each side of the plaza there is a mirrored wall facing inward as part of the enclosed arcade that surrounds the plaza and connects each vertical shaft. The mirrors reflect the clouds, sky, stars, and planets, poetically bringing them into Timeship.

The importance of informal interaction between scientists is seen in Louis Kahn's unbuilt Salk Laboratories Meeting House. Also important are Kahn's open central spaces, for example in his Exeter Library. Kahn used the term "silence" to refer to such spaces that do not have a programmed function, but draw out activities from the building's occupants over time.

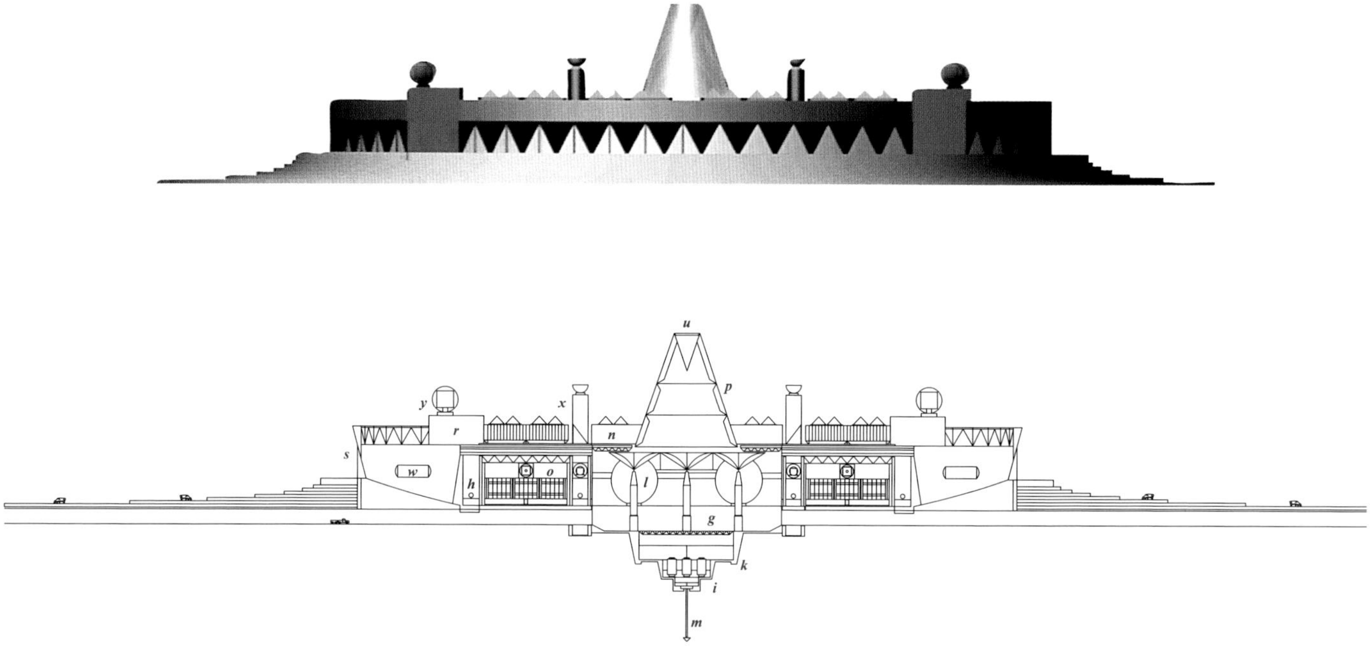

Elevation & Section

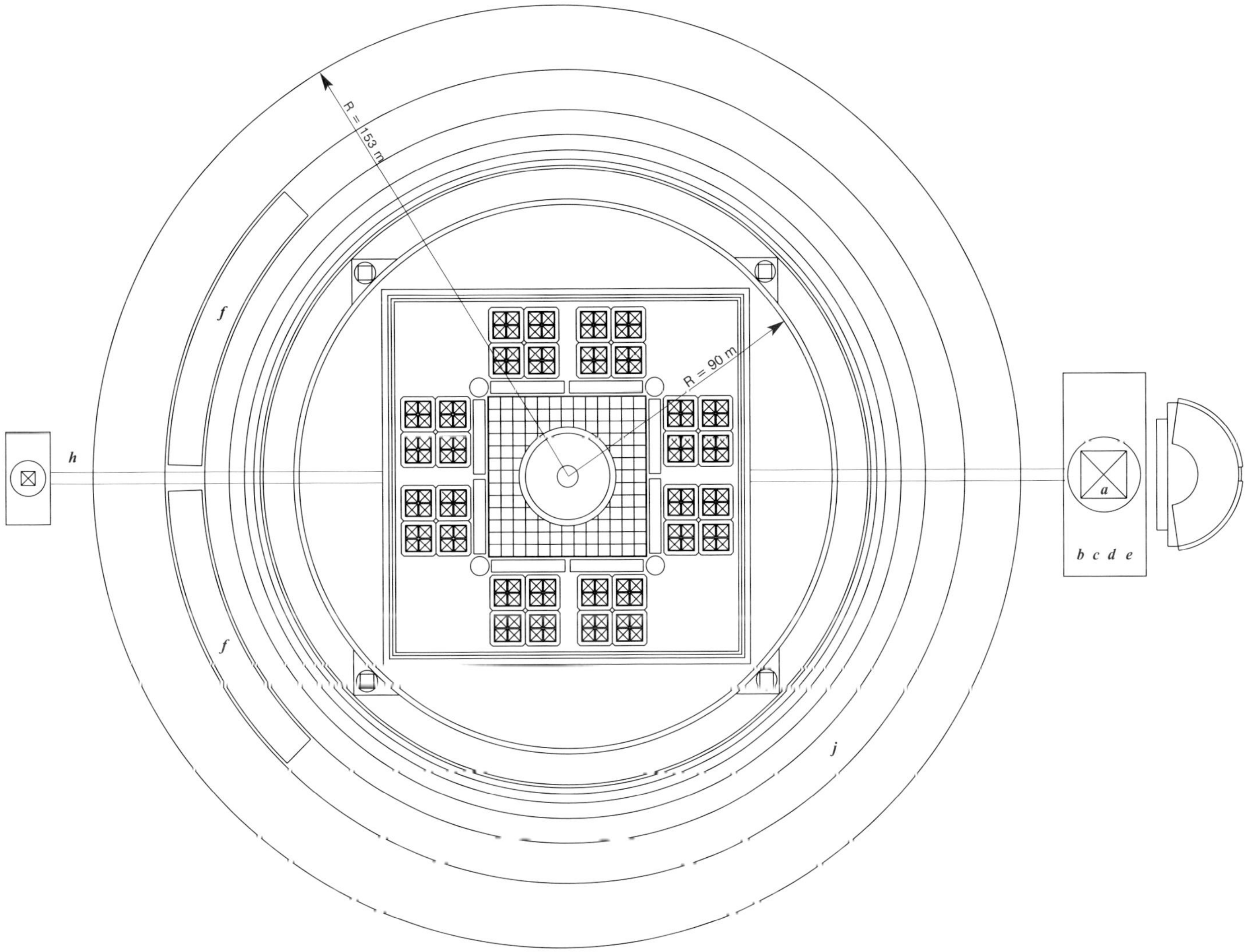

Site Plan

04 CRYOPRESERVATION

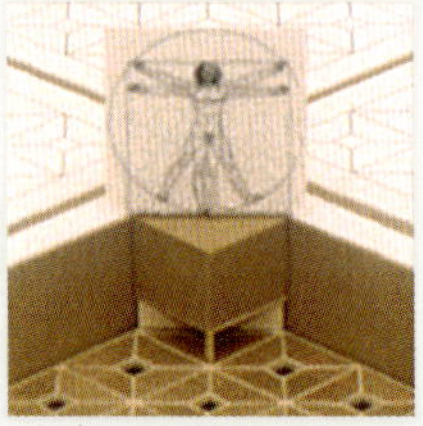

CRYOPRESERVATION

TIMESHIP'S MISSION is to conduct anti-aging research and to provide long-term, highly secure storage for biological materials and cryopreserved patients traveling to the future.

The field of human cryopreservation, or cryonics, has almost forty years of experience with the preparation and the long-term maintenance of human patients. The cryopreservation process until recently has been roughly as follows: As soon as possible after the heart of a dying patient stops, the patient is cooled down. The blood is replaced with a cryopreservation solution and the patient's temperature is then slowly lowered. The patient is then placed in a "dewar," a stainless steel vessel with a vacuum double wall like that of a thermos, which is filled with liquid nitrogen. Cooling is maintained at the temperature of -196°C by the liquid nitrogen. At this temperature, all biological and chemical activity slows down so much that there is no significant deterioration, even over hundreds of years. As long as there is a constant supply of liquid nitrogen, this temperature will be maintained.

However, despite the use of cryoprotectants, this process involves the formation of ice crystals that damage cells and tissues. Until recently, the field of cryonics has proposed to address this problem through nanotechnology as described by Ralph Merkle in his essay, "The Molecular Repair of the Brain." Nanotechnology involves the future development of tiny computerized machines, some much smaller than a human cell, and is one of the fastest growing areas of technology. Upon revival, patients would have their damaged cells repaired by nanorobots moving through their bodies. Of course, all of this is speculative and dependent on future technological developments. Other technologies, including vitrification, could make the use of nanotechnology to repair patients unnecessary.

Vitrification is the cooling of biological material to a glasslike solid state without the formation of ice crystals. Recent research breakthroughs have made possible the vitrification of patients for cryopreservation. With little or no ice crystal damage, the prospects for successful reanimation are greatly improved. This vitrification process was originally developed in the early 1980s by cryobiologist Gregory Fahy at the American Red Cross for the purpose of banking transplantable organs.

However, the vitrification process can best be applied if patients are maintained at temperatures somewhat higher than liquid nitrogen in order to prevent mechanical stresses that occur at lower temperatures. In addition, it may be desirable for different patients to be kept at slightly different temperatures, and for the temperature of a given patient to vary over time. Because of this, Timeship cryostorage must take place at varying temperatures. This must be done in a highly secure building not susceptible to disruptions such as power failures or the breakdown of control systems. Therefore, the cryopreservation technology at Timeship must be both very capable and very robust, necessitating that the technology and engineering of Timeship rival that of the International Space Station.

The vitrification process to be used at Timeship begins as soon as possible after the heart of a dying patient stops beating. Ideally, a trained cryonics team is standing by. They prepare the patient, circulate cryoprotectants, and place the patient in a rapid cool-down device to assure the survival of the patient's self and memory.

The patient is then placed in a pod in which they will remain until reanimation. There are two types of pods, "whole-body" pods and "organ" pods, which are used for organs and neuropatients. Neuropatients have chosen to have only their brains cryopreserved within their heads in anticipation of future technologies to grow new bodies around their brains. The pods are designed with multiple layers to provide both impact and thermal protection for the fragile vitrified patient during transport to long-term cryostorage.

The pod is inserted into a "transport container" that will be shipped to Timeship either by special truck or by airplane. Trucks will have refrigeration equipment, while air transport will depend on the superinsulation and minimum refrigeration capacity of the transport container.

Once Timeship is completed, pods will be placed in areas called "neighborhoods," which are grouped in three-by-three grids of nine to form "communities." There will be twelve communities arranged in a square surrounding Timeship's Prometheus Plaza, which contains backup liquid nitrogen storage tanks allowing Timeship to function for long periods even if outside supplies are interrupted. Depending on the proportions of its contents, a neighborhood can hold thousands of human organs for transplantation, as many as one hundred whole-body patients, and more than one thousand neuropatients.

The cryostorage areas at Timeship must maintain the required temperatures, be highly insulated and hermetically sealed (to prevent outside air from entering, which could lead to frost buildup). At the same time, these cryostorage areas must be accessible for bringing in new patients and organs, and the retrieval of organs for transplant. Also, those working in the facility must be protected from the cold and the remote chance of liquid nitrogen leaks.

The architect designed each neighborhood with an independent cooling system for extensive redundancy. If there is a problem with the cooling for one neighborhood, it will not affect the others, and the others can be called on as backup for the neighborhood having the disruption. The cooling system targets individual pods. The temperature in each pod is controlled by the application of forced convection of precisely temperature-controlled nitrogen gas. The cooling scheme allows for significant changes in the base temperature as required. Evaporation of liquid nitrogen provides all refrigeration requirements through the heat absorption property known as heat of vaporization. The liquid nitrogen is injected upstream of the cold circulating pump into a diffuser section, passes through additional diffusers at the pump outlet, then passes heaters to achieve precise temperature control before entering the cold volume. The superinsulation in the walls of each neighborhood and each community allows the pods to be kept cool with minimum energy expenditure. Heaters at each pod allow for pod-to-pod temperature variations.

Timeship's patient cryostorage areas, or "cold room storage," was a concept first engineered by Dr. Steven B. Harris in 1986. Proposed as an alternative to storing patients submerged in liquid nitrogen in dewars, the cold room storage approach would allow many patients to be economically maintained in a single volume. For use in Timeship, this concept was further explored in the late 1990s by Dr. Brian Wowk and Stephen Valentine.

The areas in which the staff works are separated from the areas in which there is liquid nitrogen by airlocks similar to those on a space station. Workers in areas with liquid nitrogen or subject to liquid nitrogen leaks wear a breathing apparatus for their protection and to prevent their exhalents from introducing moisture and other contaminants into the sealed environment.

Until Timeship is built, interim cryostorage of organs and patients will take place in temperature controlled volumes (TCV), containers with Timeship features. The initial design concept of the TCV was developed by cryogenics designer Michael Iarocci, physicist Brian Wowk, and architect Stephen Valentine. The TCV allows storage at temperatures compatible with vitrification, thus assuring little or no ice crystal damage.

Most people understand cryonics in part due to its frequent depiction in popular movies, and most know that the chances of eventual reanimation of cryopreserved patients are uncertain. Those who have chosen to be cryopreserved are even more knowledgeable. They know that no one who has been cryopreserved has ever been brought back, and that no experimental animal has ever been brought back. But they also know that the rate of scientific progress is accelerating. There is every reason to be optimistic that reanimation will be possible someday. (See "Scientists in Support of Cryonics" in the Appendix.)

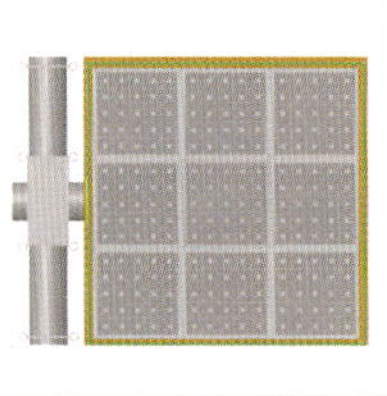

figure [*a*]

[*a*] A community with access tube. Access to the cryostorage areas must be through airtight tubes with airlocks to protect workers from LN_2 leaks, and to protect the storage area from outside moisture.

[*b*] There are twelve communities at Timeship. The hierarchy of neighborhoods and communities allows for greater security and redundancy. The inner square is for production and storage of LN_2.

[*c*] Each whole-body patient, neuropatient, and organ is stored in its own pod. A family of pods constitutes a neighborhood, the capacity of which depends on the items stored, and can number into the thousands. Nine neighborhoods make up a community.

Opposite page:
Plan of Timeship at cryostorage level.

PATIENTS AND BIOLOGICAL MATERIALS are stored in individual pods, which are placed in Timeship in a hierarchy from neighborhood to community to the entire cold storage area. One neighborhood will be made up of one hundred or more pods, depending on the ratio of whole-bodies, neuros (brains only), organs, and tissue and DNA samples. This hierarchical arrangement assures maximum safety and redundancy of the cooling systems. The cryostorage area is isolated by thermal, structural, blast, and security barriers, as well as by airlocks that prevent moisture from getting in and causing frost buildup.

figure [*b*]

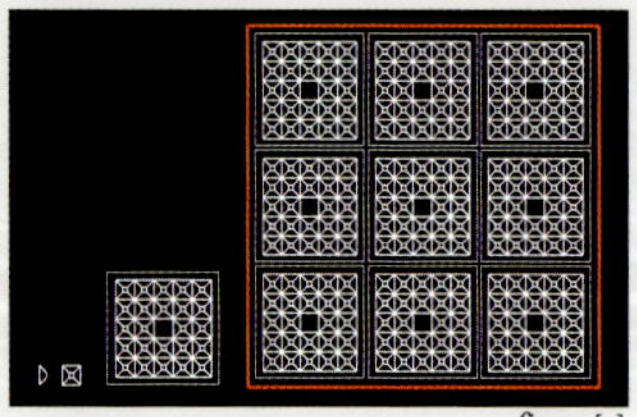

figure [*c*]

figure [*a*]

figure [*b*]

[*a*] View looking into a neighborhood with about half of the pods in place. This earlier version shows a different pod configuration than is now being proposed. The area above provides access for inserting additional pods.

[*b*] A community of nine neighborhoods. The neighborhood in the center is highlighted. Also shown is the access tube that permits entry through airlocks.

THE PODS AT TIMESHIP for whole-body patients, neuropatients, and organs for transplant are gathered into neighborhoods, which in turn are clustered into communities, with nine neighborhoods to a community. This arrangement provides maximum efficiency in terms of space and energy consumption. The tighter the pods are packed, the less the surface area in relation to volume, and therefore the less the thermal energy transfer, resulting in less refrigeration needed to maintain the cold temperatures. Despite being a very large building with demanding requirements, Timeship will be highly energy efficient and able to store biological materials economically.

Efficient packing also leads to structural efficiencies, since spans can be reduced. The structural system at Timeship must provide security against numerous natural and man-made threats and last for hundreds of years.

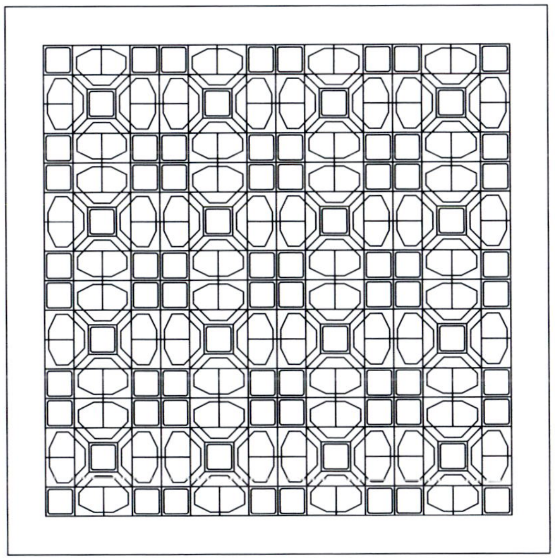

figure [*a*]

figure [*b*]

CLOSE PACKING OF THE PODS in the neighborhoods and communities at Timeship is necessary for maximum energy efficiency. One could use hexagons as in a beehive, which has certain advantages, but patient storage requires a balancing of several factors. The final choice of packing geometry must consider the size of the pods in order to accommodate the greatest range of patients; the ratio of whole-body patients to neuropatients and organs; the interior volume to surface area ratio, which affects heat transfer; the packing properties of various geometries; and the configuration of the neighborhoods and communities in relation to the design of the building as a whole. Hexagonal packing would have led to an increased surface area, and therefore increased heat transfer and loss of efficiency.

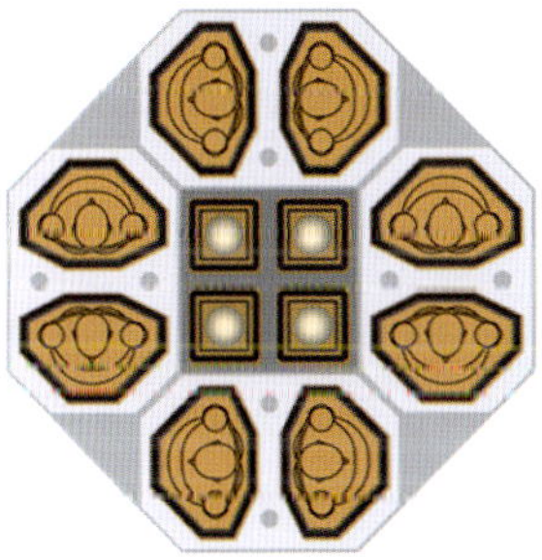

figure [*c*]

[*a*] & [*b*] Studies of various packing geometries that were considered during the early design phase. One shows beehive hexagons that were rejected. Another has spaces for control devices at the center of each cluster of four whole-bodies, but these devices are not needed in the current design.

[*c*] Whole-body patients symmetrically configured around organ pods.

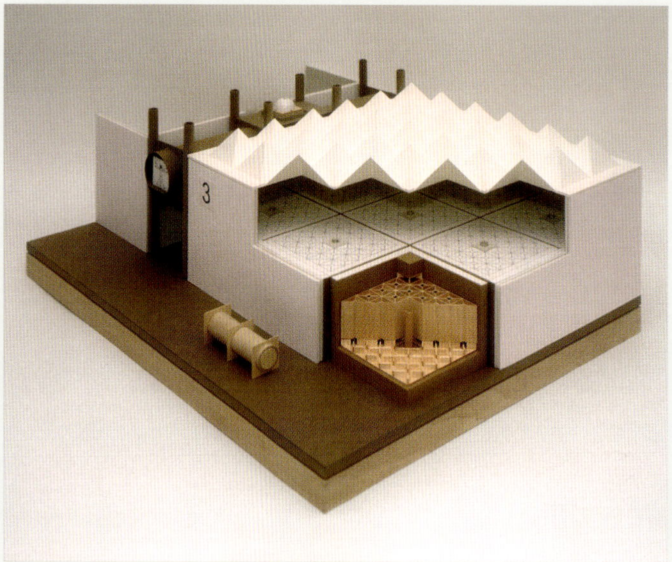

figure [*a*]

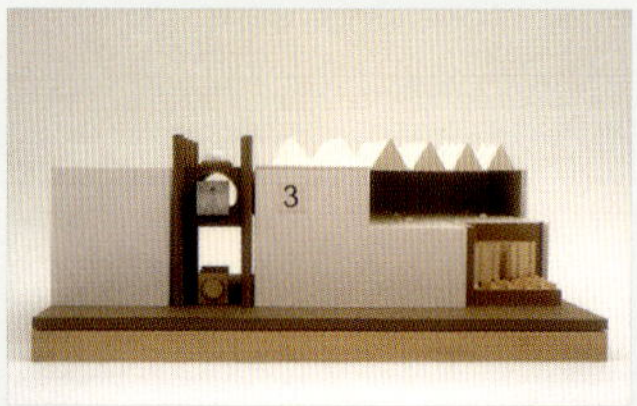

figure [*b*]

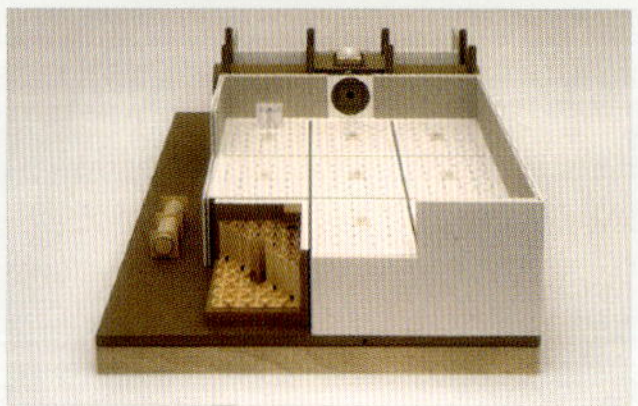

figure [*c*]

THE CRYOSTORAGE AREA at the heart of Timeship is made up of twelve communities. The communities provide the ultimate in security with multiple layers of protection, including superinsulated evacuated walls and a blast-resistant structural truss system for the upper enclosure. There is structural and mechanical redundancy throughout.

New patients already in their pods will be brought into the upper level of a community through an airlock to protect against cold escaping and moisture getting in. Working in this upper level, Timeship staff will lower the patient pod into the cold storage volume area.

Although the areas around communities will not be overly cold because of the superinsulation, workers entering these areas for infrequent maintenance will have respirators in case of a catastrophic LN_2 leak. LN_2 is not toxic in itself, but as it boils the evaporating gas rapidly expands and displaces oxygen.

[*a*], [*b*] & [*c*] A community of nine neighborhoods, an access tube, and an LN_2 tank. The small pyramids at the top represent the structural truss system of the blast-resistant upper enclosure that provides a high degree of security. The open area under the upper enclosure is the patient transfer volume for maneuvering patients to be lowered into the suspension matrix grid. The corner that is further cut away is a neighborhood showing an earlier suspension matrix grid configuration that holds patient pods. Although only one is shown, there will be numerous LN_2 storage tanks.

figure [*a*]

[*a*] When Timeship workers move through the cold storage areas, they will remain inside access tubes that protect the environment as much as the staff. When these tubes penetrate into patient cryostorage areas, airlocks must be used both for thermal isolation and to prevent the moisture in normal air from getting into these areas and causing frost that would be difficult to remove.

[*b*] Airlock from a space station, similar to those that will be used on the access tubes at Timeship.

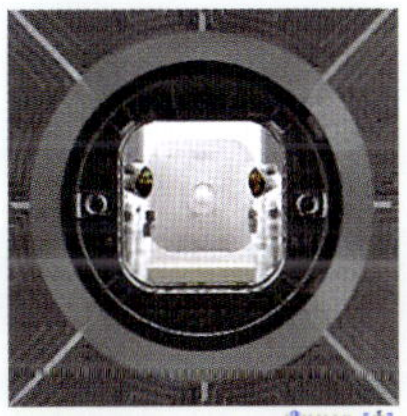

figure [*b*]

COMPARISONS BETWEEN TIMESHIP and a spacecraft are appropriate in many ways: both carry human travelers, one through space and the other through time. Both must protect their passengers from a hostile and unforgiving environment, one from the vacuum of space, the other from the heat of everyday temperatures and from natural and human threats. And just as the space station must separate its inhabitants from the vacuum of space, Timeship must separate its workers from the low temperatures and unbreathable nitrogen atmosphere of its cold storage areas. To facilitate this, the building is divided into two interlocking realms: a safe world and a cold world. An elaborate system of airtight tubes with airlocks facilitates movement of the staff in the cold storage area.

An image in the illustration above shows Leonardo da Vinci's Vitruvian Man. This human placed in a diagram of a squared circle is used here to symbolize the human presence in this high-tech environment and in our high-tech world. With Leonardo and the Renaissance we see the emergence of humanism, the notion that we can understand and improve our world and ourselves through our human powers of observation and reason.

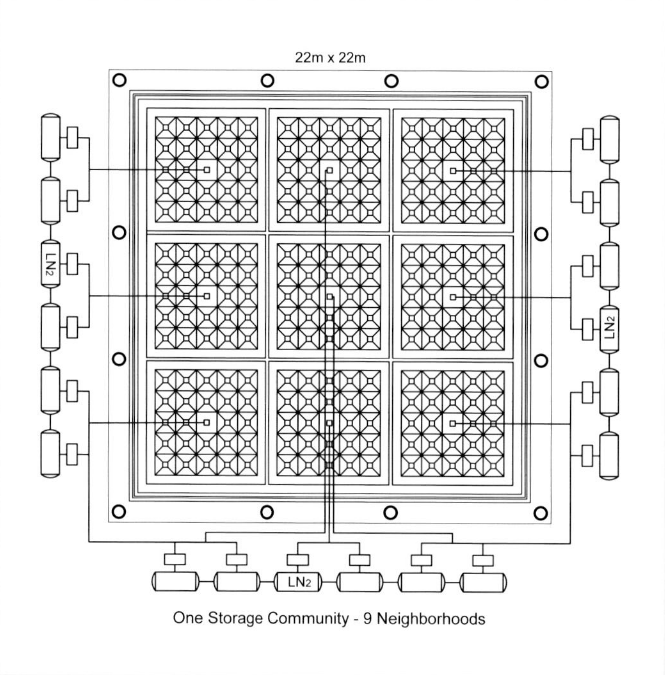

figure [*a*]

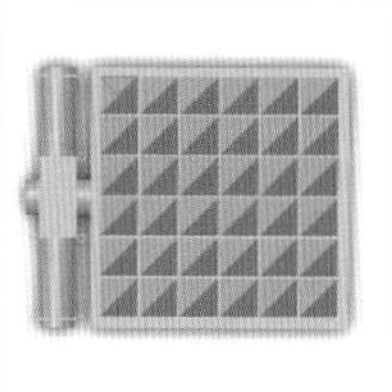

figure [*b*]

[*a*] Diagram of the LN_2 storage and distribution for one community. Surrounding each community in the corridors there are pairs of local LN_2 supply tanks that provide the refrigeration for each neighborhood. The corridors will also house the systems that control the temperatures. Each neighborhood is supported by an independent cooling system and there are multiple redundancies, with one neighborhood's systems able to back up another's.

[*b*] Community with truss system upper enclosure and the access tube for staff circulation and bringing in new patients.

COOLING AT TIMESHIP must be reliable for possibly hundreds of years, maintaining temperatures within a couple of degrees and not allowing its precious cargo to "thaw." In addition, it may be desirable for some patients to be kept at different temperatures, and it also may be desirable to vary the temperature of a given patient over time. Therefore, for cooling, the architect chose evaporating LN_2, which extracts heat from its surroundings, minimizes moving parts, and eliminates compressors for maximum reliability. LN_2 is also very economical. The LN_2 is injected upstream of the cold circulating pump into a diffuser section. It then passes through additional diffusers at the pump outlet, passes heaters to achieve precise temperature controls, and then enters the cold volume. The temperature at each pod is controlled by the forced convection of precisely temperature-controlled nitrogen gas and localized heaters.

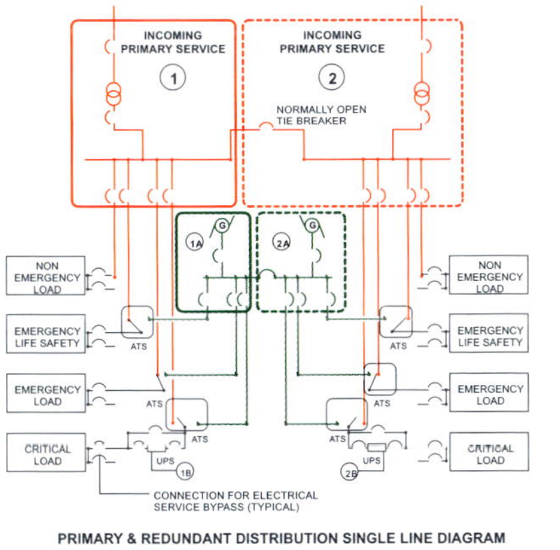

figure [*a*]

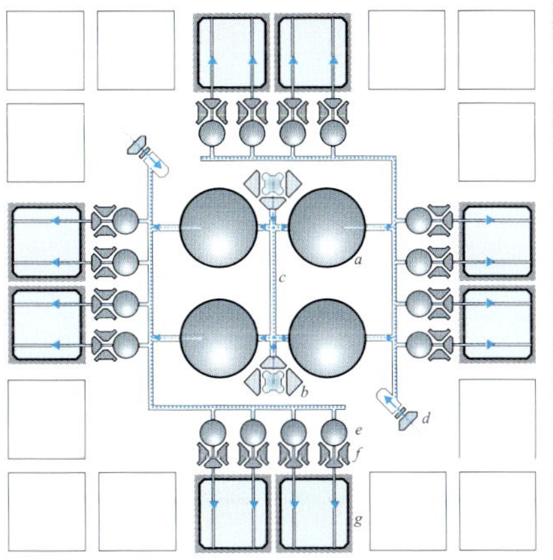

figure [*b*]

[*a*] An electrical scheme for the patient cryo-storage area of Timeship showing a high degree of redundancy to assure uninterrupted temperature regulation. Electrical systems will use the highest grades of materials, but be of basically simple design to assure reliability.

[*b*] Timeship will keep a backup supply of LN_2 should deliveries be interrupted. It will also be able to make its own LN_2. This diagram shows the distribution of LN_2 within Timeship and the extensive redundancy of supply to each area.

[*c*] LN_2 backup storage tanks above Prometheus Plaza at the center of Timeship.

figure [*c*]

TIMESHIP'S "TIME TRAVELERS" are dependent on the reliability of the cooling systems and their controls. Because of this dependence, architect Stephen Valentine retained some of the world's leading mechanical engineers to design Timeship's systems. The charge Valentine gave the team was as follows: "Imagine a block of ice with your name carved in it in bold letters like the date we might see in a block of stone in the Roman Forum. That name carved in the ice is your identity. Should the block even partially melt and your name blur, your identity will disappear, even if the block is immediately refrozen. Your assignment is to create a cooling system of such reliability that your name will remain crisp and readable for up to hundreds of years."

Reliability of the cooling systems at Timeship is achieved through a combination of simplicity of design, high-quality materials and components, and a high degree of redundancy.

To enhance reliability, the cooling comes from the evaporation of LN_2 rather than from compressors. This greatly reduces the number of moving parts. LN_2 is inexpensive and readily available, and on-site refrigeration equipment will generate LN_2 should there be an interruption of outside supplies.

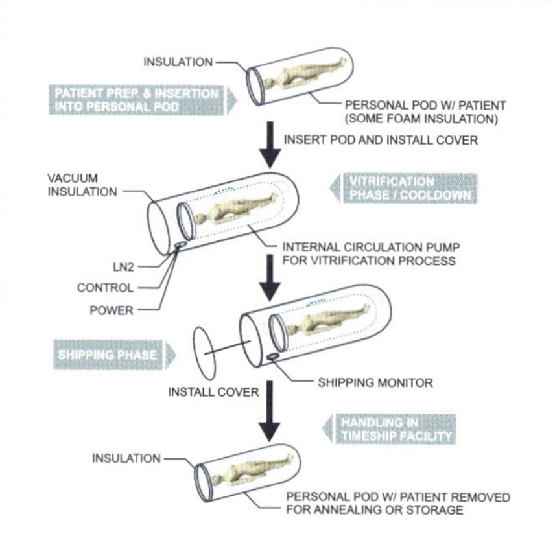

figure [*a*]

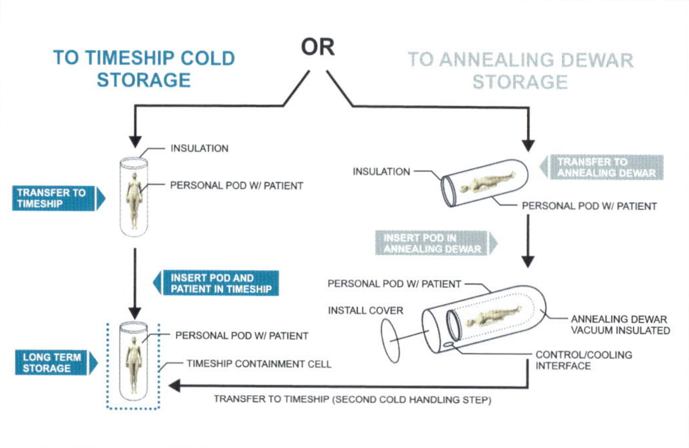

figure [*b*]

[*a*] Containers used for cryopreparation and preparing the patient for shipping. Teams with equipment are ready to respond at a moment's notice.

[*b*] Two alternatives: First, the patient goes directly to Timeship, second, to an interim temperature controlled volume (TCV) before Timeship is operational.

[*c*] Mobile rapid cool down device. Once a patient is officially declared dead, cool-down must be done quickly to preserve the patient's "self" and memory in the brain.

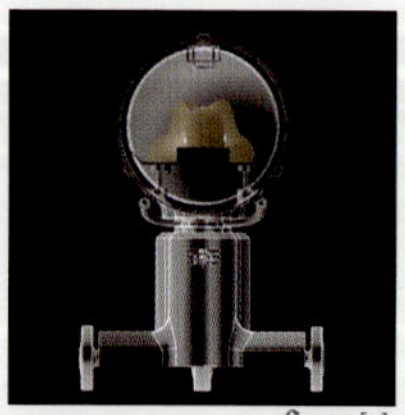

figure [*c*]

PATIENT CRYOPREPARATION, shipping, and cryostorage are complex and technically demanding processes. The patient cannot simply be immersed in LN_2, as doing so would lead to the inner volumes of the patient not cooling fast enough and cracks from uneven cooling. Additionally, LN_2 is not at the ideal temperature for vitrification.

The process begins with a specially trained team that waits until the patient's heart stops. The team uses modern medical equipment, including a heart-lung machine, to keep blood flowing to the brain while infusing the patient with ice blocking substances and inducing cooling. The patient is then placed in a personal pod and rapid cooling is completed. The entire process resembles a major surgical procedure. The pod is then placed into a transport container for shipping. The patient will remain in the same pod until eventual reanimation.

Until Timeship is completed, patients will be placed in interim long-term storage temperature controlled volumes (TCV). After Timeship is completed, they will be transferred to the facility and placed into secure long-term cryostorage.

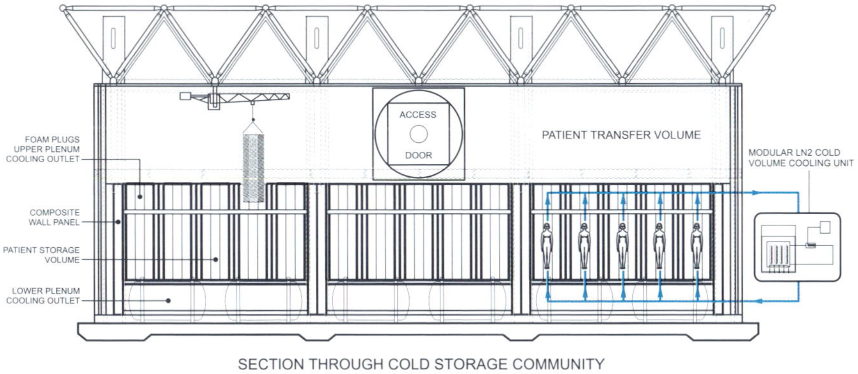

figure [*a*]

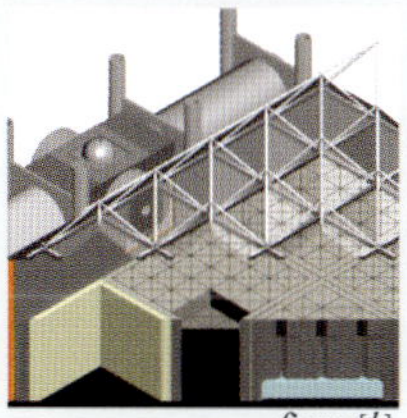

figure [*b*]

[*a*] Diagram showing a community, the airlock access door, the patient cryostorage volume with five patients shown, and the patient transfer volume above for maneuvering the patient. The patient transfer volume and the individual patient storage units are separated by foam plugs. The crane is lowering a patient pod. To the right is one of many cooling units.

[*b*] Cutaway isometric showing a community.

[*c*] Section through a community. The blue triangles are the structural truss system of the blast-resistant upper enclosure. The gray below is an access tube for moving patients. The orange is the patient cryostorage volume. The red tanks, two per neighborhood, hold LN_2 for cooling.

THE WALLS OF THE NEIGHBORHOODS and communities are superinsulated and lined with welded stainless steel to make them airtight in order to maintain vacuum insulation and prevent infiltration of moisture. Cooling and reheating are applied to the cold storage area as a whole and to each individual unit, as temperatures may have to vary slightly over time for some patients.

Technicians bring pods containing new patients, or organs for future transplant, along the access tubes, through an airlock, and into the patient transfer volume. The pod is lifted by a crane attached to the ceiling structure, maneuvered over and then lowered into the patient cryostorage volume matrix. Patients will then await eventual removal and reanimation.

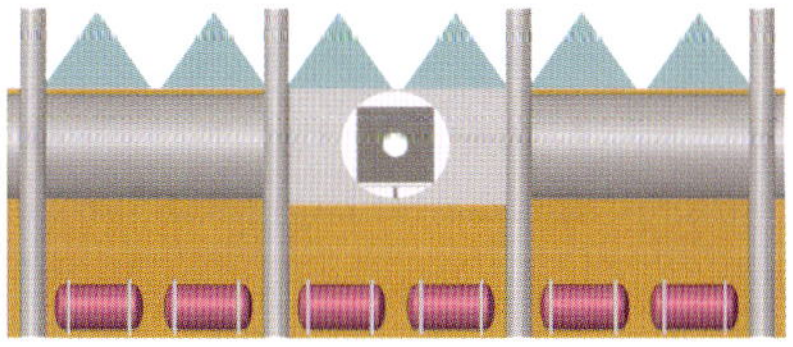

figure [*c*]

INITIATED IN 1958 and completed in 1963, Project Mercury was the first man-in-space program of the United States. The goal of the Timeship pod is closer to that of Project Mercury than that of an Egyptian sarcophagus—Timeship seeks to bring its patients back safely. It also shares with space ventures a need for precision.

The Timeship design works outward from the pods. However, because the pods will be put into use before Timeship is completed, they must work both in the final building and in interim storage temperature controlled volumes (TCV) that will cluster together a combination of neuro and whole-body patient pods inside a dewar.

figure [*a*]

[*a*] The small one-man Mercury space capsule designed by Max Faget and his team had 1.7 cubic meters inside.

[*b*] King Tut's sarcophagus with multiple layers protected him on the journey to the afterlife.

[*c*] Multiple layers of a whole-body pod protect patients on their journey to the future.

Opposite page: Cutaway view of a TCV with four whole-body patient pods.

figure [*b*]

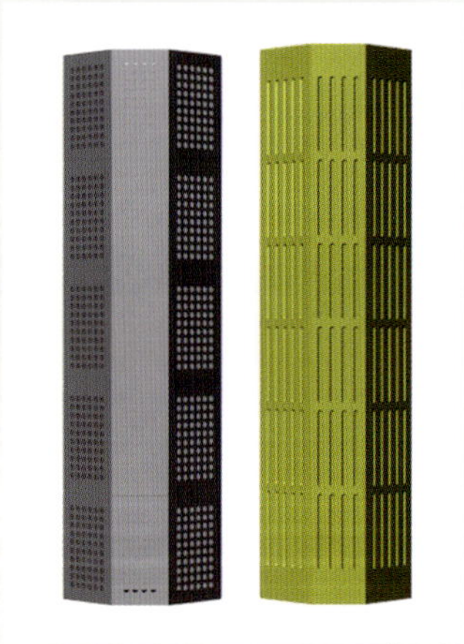

figure [*c*]

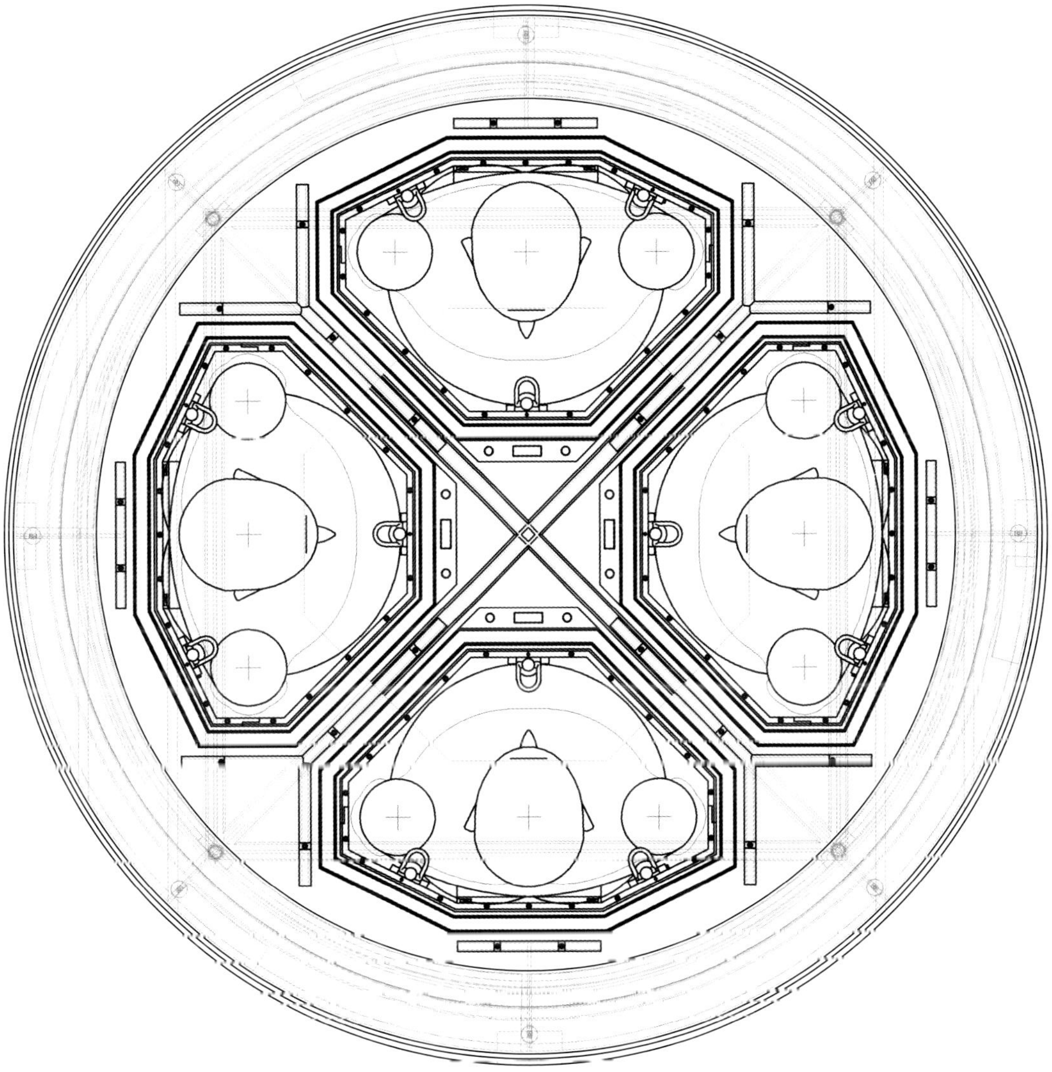

ARCHITECTURAL DESIGN INVOLVES numerous decisions based on extensive research. One example was establishing the size for the whole-body patient pod to be used at Timeship. People come in a wide range of sizes, and existing ergonometric and anthropometric standards had proved to be inadequate because of frequent errors, necessitating that the architect do in-depth research. The measurements used for Timeship had to account for the fact that people on average grow larger as they age, and that they spread out when they are lying down. The architect determined the range of body sizes in the world today, finally choosing what is called a 99-percentile body size. This means that 99 percent of all people will be able to fit in the whole-body patient pod designed for Timeship. Those who are larger cannot be accommodated in whole-body pods, but they may choose neuropreservation.

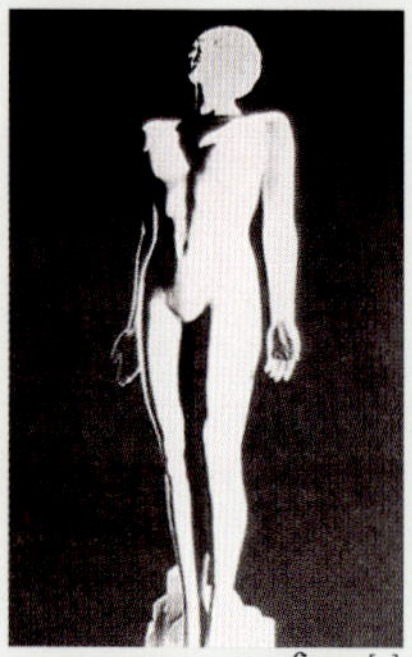

figure [*a*]

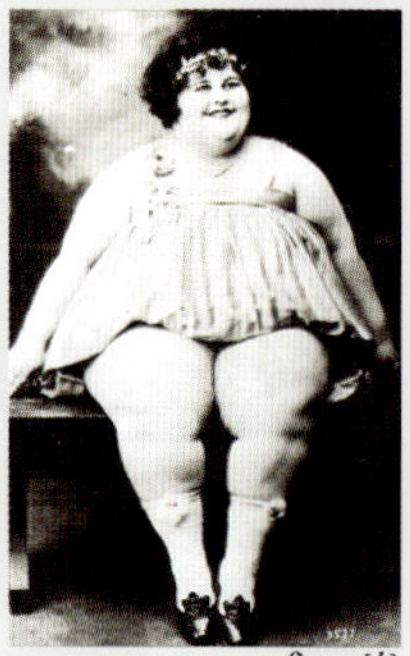

figure [*b*]

[*a*] & [*b*] There is a wide range of body sizes and shapes. The pod is not designed for the "average" or "typical" person, but for 99 percent of people.

[*c*] Configuration of the whole-body patient pod is based on extensive ergonometric and anthropometric studies, and on packing geometries.

[*d*] Detailed dimensions of the 99-percentile person used to design the pod.

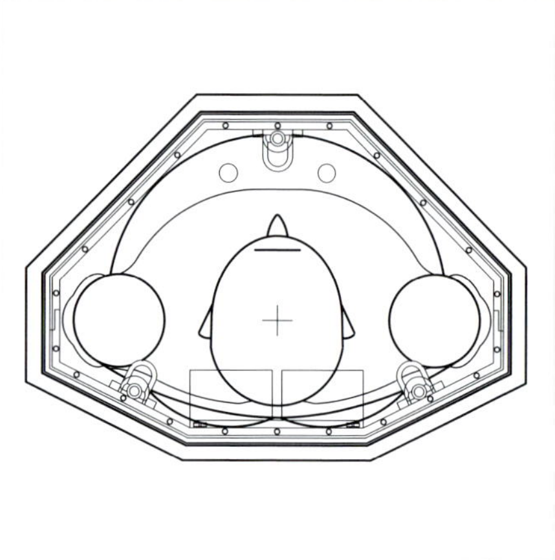

figure [*c*]

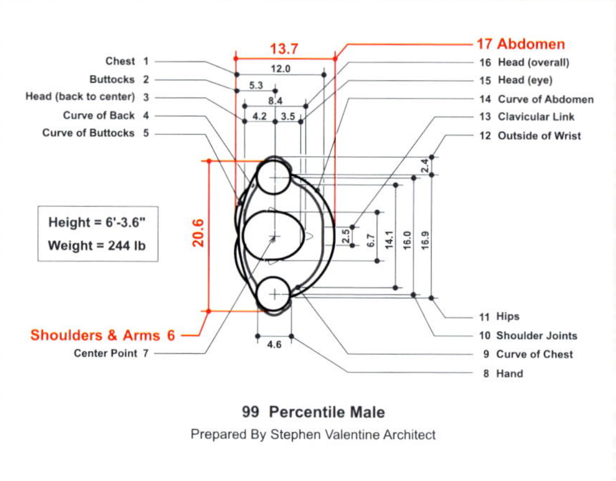

figure [*d*]

A TIMESHIP PATIENT might begin his or her journey from almost any place in the world. After the end of one's "first life cycle," the initial phases of cryopreparation take place nearby and the patient is placed in a pod which some will arrange to be kept nearby. The pod is then put into a transport container that will be shipped to Timeship either by special truck or by airplane. Trucks will have refrigeration equipment, and the trip need not be rushed. Shipments by air will have to depend on the superinsulation and the minimum refrigeration capacity of the transport container, which will permit trips of up to twelve hours.

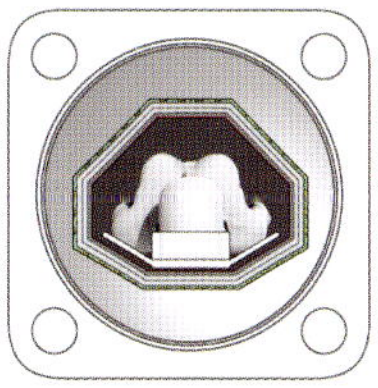

figure [*a*]

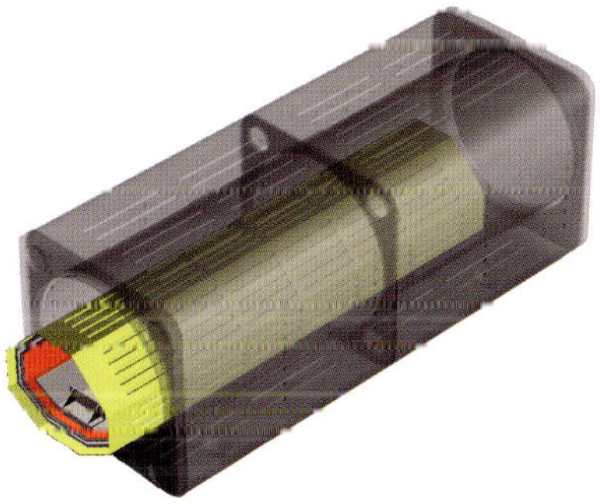

figure [*b*]

[*a*] Cutaway view of a transport container. A suspension system with shock absorption and superinsulation to insure the patient's safe journey to Timeship or to interim storage (TCV).

[*b*] Isometric view with the cover removed to show a whole-body patient pod inserted into the transport container.

[*c*] Transport container. The yellow form is the patient pod.

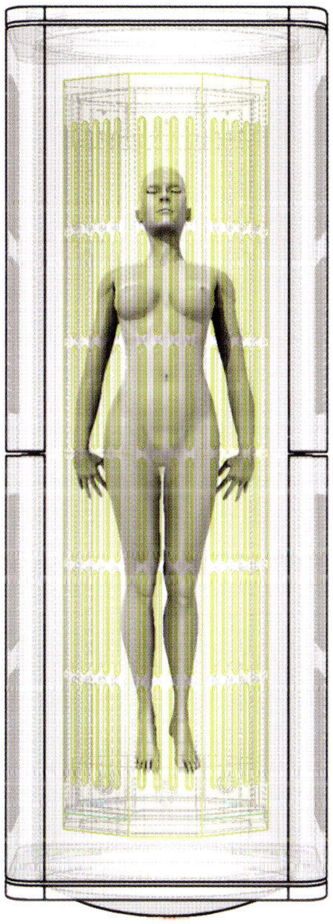

figure [*c*]

figure [*a*]

[*a*] The individual components of a TCV. Minimal heat transfer will be achieved through double-wall construction, vacuum, and insulation. A superinsulated and vacuum wall is located in the annular space between vessels in order to minimize heat flow into the TCV. This vacuum space is also common with the other vacuum spaces at the facility so they can be periodically evacuated together as necessary. The heat flow of the TCV was modeled analytically using the electrical resistance analogy approach.

[*b*] Looking into a TCV. The multiple layers of insulation are apparent.

figure [*b*]

UNTIL TIMESHIP IS BUILT, both patients and organs will be placed in temperature controlled volumes (TCV) for interim cryostorage. In the TCV, a vertical open mouth containment dewar with foam cover is used to house a second inner dewar. A conductive aluminum liner within the interior dewar enables control of the spatial temperature. Both vessels have vacuum insulation and superinsulation to minimize heat transfer, and the vessel walls' vacuum spaces are common for easy re-evacuation from outside the dewar if needed. The bottom few inches of the outer dewar contain LN_2. Cold conductors, sensors, and heaters complete the scheme for precise temperature control.

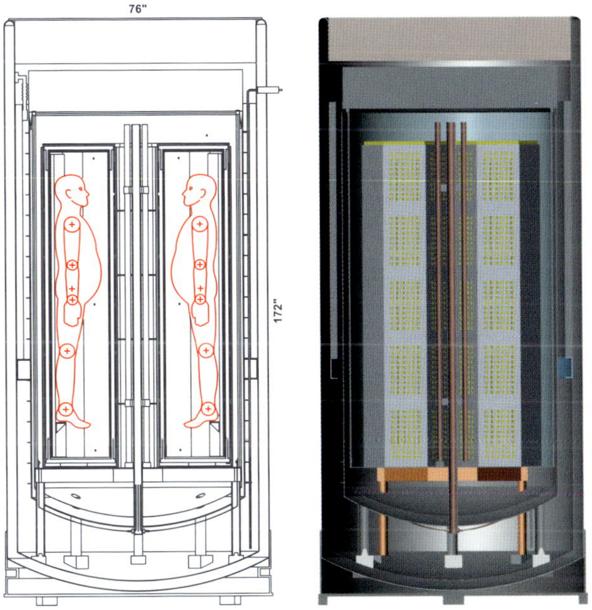

figure [*a*]

MAINTAINING LN2 TEMPERATURES is simple: LN_2 is poured into a dewar and is continually replaced as it boils off. However, because cryonic vitrification requires temperatures somewhat higher than LN_2 and perhaps even slightly different temperatures for each pod, a temperature control system is needed. A mostly passive cooling system was adopted to reduce the possibility of failure from moving parts. At the bottom of the TCV, connected to the highly heat-conductive aluminum liner, heat conductors are positioned so that they can be immersed in or removed from an LN_2 bath. Adjusting these conductors via control cables changes the thermal resistance between the TCV bottom surface and LN_2 pool, and changes the heat transferred to the LN_2, thereby allowing control of the temperature of the TCV. Small resistive heaters are located in the TCV lower area to "trim" and more finely control the temperature as may be required.

figure [*b*]

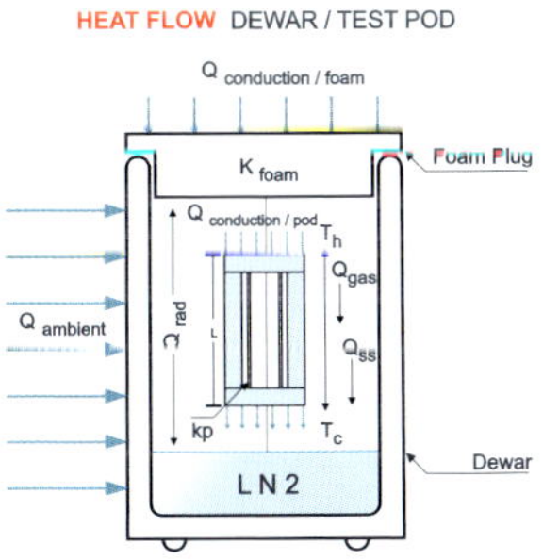

figure [*c*]

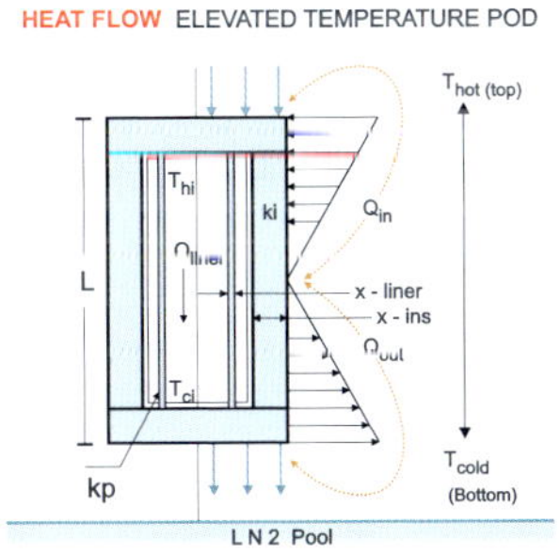

figure [*d*]

[*a*] Vertical sections through a TCV.

[*b*] Controls for fine temperature tuning.

[*c*] Heat flow in a pod is controlled by a conductive liner in an insulated shell. LN_2 at the bottom does not touch the inner liner.

[*d*] Heat flow analysis.

[*a*] The parts of the whole-body patient pod: from the left, the yellow tube is the protective enclosure (with its top and bottom). Within this tube goes the stainless steel vacuum-insulated volume and the aluminum heat-conductive liner to assure that the whole-body patient is kept at an even temperature. The patient capsule is placed inside the liner, and inside that the patient cradle is inserted (shown extended). The last element to the far right is the sleeve that is part of the TCV into which the whole-body patient pod is placed for interim storage.

[*b*] Structural embossing on protective enclosure.

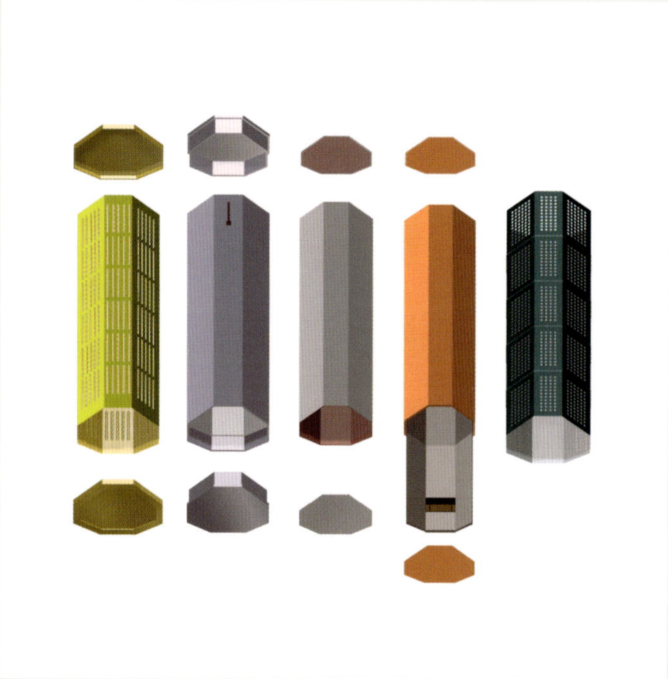

figure [*a*]

figure [*b*]

THE EGYPTIANS USED multiple layers with spiritual functions for the capsule that took the dead to the afterlife. At Timeship, the multiple layers of the pod serve practical functions. The pod is the immediate container for the patient, providing the first line of defense against damage, time, and heat. In addition, it is the basic module of Timeship. The entire building is designed from the pod outward to the neighborhood, to the community, to the cold storage area, and to the building as a whole.

The pod must provide both impact and thermal protection for the fragile vitrified patient during transport to long-term patient storage. Toward this objective, a complex design was developed with multiple layers. Each layer has a protective or thermal role in bringing the patient safely to the future for eventual reanimation. The same pod will be used in both the interim TCV and the final building.

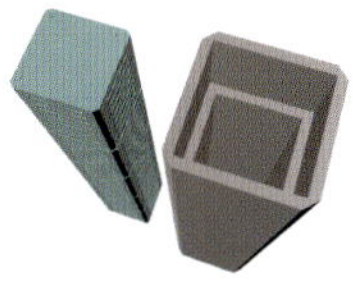

figure [*a*]

[*a*] Five stacked organ pods next to their external liner.

[*b*] View looking into an organ pod showing the assembled elements.

[*c*] Parts of the organ pod: The sphere at the center represents an organ such as a kidney, and the various layers are for protection, g-force resistance, and thermal insulation.

[*d*] Diagram showing an organ pod's g-force protection.

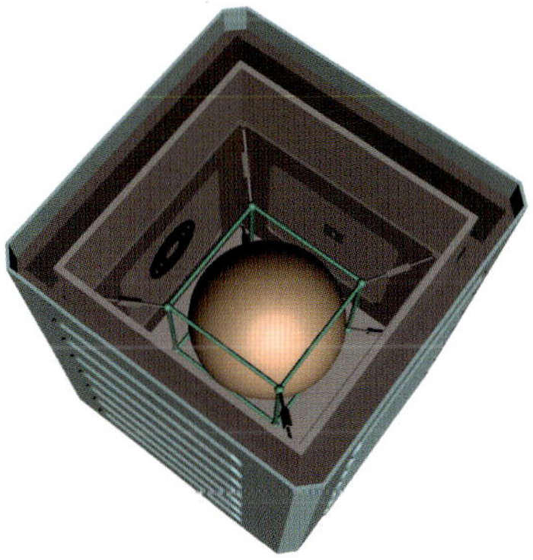

figure [*b*]

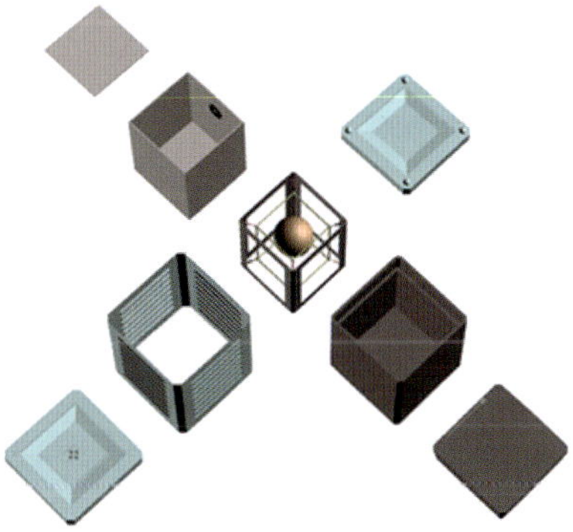

figure [*c*]

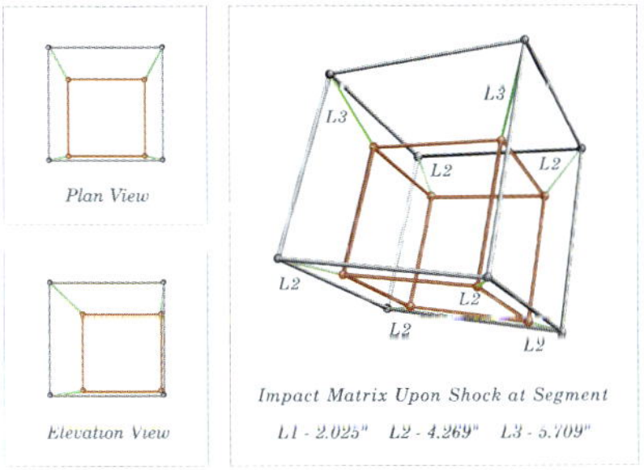

figure [*d*]

TIMESHIP WILL CRYOPRESERVE whole-body patients, neuropatients (many patients choose to have just their brains preserved), organs, and other biological material. An organ pod is used for all cases other than whole-body suspensions.

Laptop computers are engineered as much as possible to withstand the g-forces of a minimal drop. The "self" and memories of Timeship's patients are stored in their brains, and in their solid vitrified states they are vulnerable to even minor bumps. A careful g-force analysis was done and a unique friction spring suspension armature system was designed to protect an organ against such accidents.

05 BUILDING TECHNOLOGY

BUILDING TECHNOLOGY

TIMESHIP WILL BE one of the most demanding and complex buildings ever built, on a par with the International Space Station in its mission, its challenges, its complexity, its needs for reliability and redundancy, and its requirements for error-free design and construction. Timeship must keep its precious cargo of cryopreserved DNA, biological samples, organs for transplant, and patients traveling to the future safe, in some cases for over one hundred years—all with no interruption of perfectly constant temperatures, no disruptions from the outside, and no systems failures. All of Timeship's building technologies, including structural, mechanical, and security systems, must be dedicated to this mission.

Timeship must embody and convey permanence. We live in what is often called a "disposable society," but with greatly extended life spans, we will begin to adopt new attitudes that embrace timelessness, and Timeship, through its symbolic expression, will play a role in fostering these new attitudes. While we are all familiar with the impermanence of most consumer goods, even buildings deteriorate, and Timeship must stand in defiance of that tendency as well by operating without interruption for up to hundreds of years.

Due to the highly vulnerable nature of Timeship's cargo, security was the central focus of Valentine's design effort from the very beginning. Because of this concern, he retained some of the world's foremost security consultants to work on the project and be involved with every aspect of the design. Security might be defined as the opportunity for Timeship's mission to be uninterrupted in the protection and preservation of critical organisms and life forms and their support systems for future generations. Put simply, this means nothing should go wrong, and if it does, no irreparable damage will be done.

Interestingly, security begins not with thick walls and iris scanners, but with public relations. The society as a whole—and particularly Timeship's immediate neighbors—should view the enterprise favorably. If neighbors are supportive, they will be a first line of warning. So the security analysis begins with the very definition of the mission, includes site selection and relationships with immediate and regional neighbors, and then extends to all aspects of the design. With regard to the building itself, Valentine repeatedly stated that his security designers should begin by assuming that only sixteenth-century technology was available. In other words, the building should be inherently secure by nature of its basic layout even if there is an electrical failure. State-of-the-art electronic devices should be added only after the initial design is secure. One example of this is the landscape design at the perimeter of the building. The series of stepped rings surrounding Timeship not only have symbolic meanings, but also provide standoff protection against potential truck bombs. Massive walls and roofs provide the next line of security.

Other design features for security include a single highly controlled underground entrance, ventilating systems designed to protect against the introduction of toxic agents, extensive electronic monitoring, and electronics shielded against electromagnetic pulse (EMP) and other threats.

The structural systems of Timeship must not only protect against blast and other potential attacks, but must also have a long life span with low maintenance. Since structural materials and any support system, no matter how carefully designed and installed, might someday succumb to hidden defects, improper installation, or just wear and tear, the Timeship structure must be visible, testable, and capable of having components easily swapped out without disrupting operations.

Due to the possibility of some form of attack over its life span, Timeship must be designed to resist blasts. The first level of defense is to prevent an attack.

The next is to create "standoff" so that a blast happens far enough away to do minimal damage. The final level is to use materials and designs that will sustain the least amount of damage. The structure will be designed to prevent progressive collapse and limit excessive damage to vital areas and critical systems. These features typically involve thick slabs and additional reinforcement to withstand abnormal loading patterns. The use of moment-frame connections will improve the structural redundancy, the encasement of select columns will provide additional capacity and ductility, and the use of hardened walls will confine interior explosions and limit the extent of any damage.

While these measures will limit blast damage to Timeship, the detonation of an internal bomb at any location could result in extensive damage and injury. This fact highlights the importance of an effective screening procedure to prevent a hand-carried bomb from entering the facility.

New materials, particularly composites, and emerging new technologies may offer the possibility of marked improvements in performance in the future. The Timeship design must be sufficiently adaptable to allow for alterations and adjustments to take advantage of upgrades made possible by such developments.

One of the most serious potential threats to Timeship is from earthquakes, which is addressed first by locating in a zone of least occurrence. Notwithstanding, all components of the mechanical and electrical equipment, as well as distribution systems, will be designed to resist seismic forces. This program requirement will protect against any seismic activity that may occur even in zones that have traditionally been considered as zero seismic activity areas. Additionally, in many ways large blasts produce loads that mimic seismic activity. Thus, structures and mechanical systems designed to resist one will resist the other.

Architecture gives form and function to a building. Structure insures its stability. These two design disciplines can produce beautiful, symbolic, functional, safe, and enduring structures. However, for buildings to function, mechanical, electrical, hydraulic, and life safety systems are equally important. The electromechanical systems required for Timeship include the heating, air conditioning, ventilating, electrical power, lighting, low-voltage communications and data systems, patient cooling controls, water supply, plumbing, drainage, and fire protection systems, including both detection and suppression. While an ultra-secure facility such as Timeship must consider many factors of longevity and safety, fire statistics for even the most modern and fire-safe projects suggest that in the life span of this facility the most likely peril it will face is fire. Consequently, every effort will be taken to protect against fire, including a fire suppression system that will take advantage of the abundant LN_2 on site.

While the structural systems of many buildings can last hundreds of years, the mechanical and electrical systems cannot. Therefore, the mechanical and electrical systems for Timeship will be designed for easy removal and replacement when required.

In order to help its longevity and to minimize its impact on the environment, Timeship will be a sustainable "green" building. It will use safe, environmentally friendly materials, provide its workers with attractive and toxin-free workplaces, be highly energy efficient, and use renewable energy sources where possible.

Due to the fact that Timeship might be operational for more than one hundred years, and during that time must suffer no interruptions in its operations, specific attention has been paid to energy. First, due to insulation and careful design, Timeship will be highly efficient, requiring very little energy input to maintain constant cryopreservation temperatures. In observing changing energy technologies over the past hundred years, we can anticipate that there will be even greater changes over the next hundred years. Timeship will initially be on the grid, accessing several redundant sources of energy, and it will have its own means of auxiliary energy generation. And Timeship will use alternative energy sources, such as geothermal, solar, and fuel cells where appropriate depending on the depletion of current fuels and on developing technologies.

figure [*a*]

[*a*] The first inspiration for the security of Timeship was castles. With their impregnable locations, thick walls, watchtowers, and moats, castles inspire us from childhood as places of security. The basic principles of castles still have much to teach us and are incorporated in Timeship.

[*b*] & [*c*] World War II bunkers also influenced Timeship's approach to the security design.

figure [*b*]

TIMESHIP WILL BE A "FORT KNOX" for the storage of biological materials, protecting its precious cargo through hundreds of years of "travel into the future." During this trip the passengers will be in a highly vulnerable condition. For example, a change in temperature of only a few degrees could ruin the trip. The conditions that might threaten Timeship are many—both likely, such as fire, which most buildings will suffer at some point, and unlikely but still potentially catastrophic, such as a collapse of the social order. Security is a top priority at Timeship, and some of the world's leading security consultants were engaged on the project and worked with all of the designers at every level.

The architect has incorporated the basic principles of security in Timeship: good public relations and a favorable perception by the general public so that people will not feel hostile to Timeship; friendly neighbors to provide warnings and assistance; standoff terracing to keep threats at a distance; design, including limited access, strong structure, and survivability; reliable personnel; and only after these have been addressed, state-of-the-art technology.

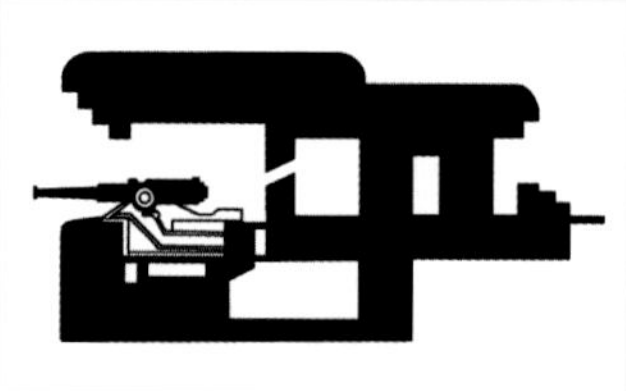
figure [*c*]

figure [a]

[*a*] Viewed from the ground, Timeship appears friendly in its environment, but also secure. One cannot approach too close due to the surrounding stepped rings of earth. The outer wall will be patterned, but made of impregnable reinforced concrete. Timeship will be built to allow the substitution or addition of new materials such as high-strength composites, as they become available, to reinforce the construction.

[*b*] Fortifications also provided models for the design of Timeship.

WHEN THINKING OF SECURITY TODAY, we often think of video monitors, iris scanners, and other high-tech equipment. These are important, but secondary. Security must be in every aspect of the planning process from the beginning. To inspire his security consultants, the architect asked that they begin by thinking of premodern construction, perhaps sixteenth-century fortifications. In this approach, security is a part of the basic architectural organization: entry, circulation, visibility, and locations of the spaces are all approached from a security point of view. Once that is done, then the high-tech devices can be added. But should all of the electronics be disabled, for example by an electromagnetic pulse, the building must still be secure.

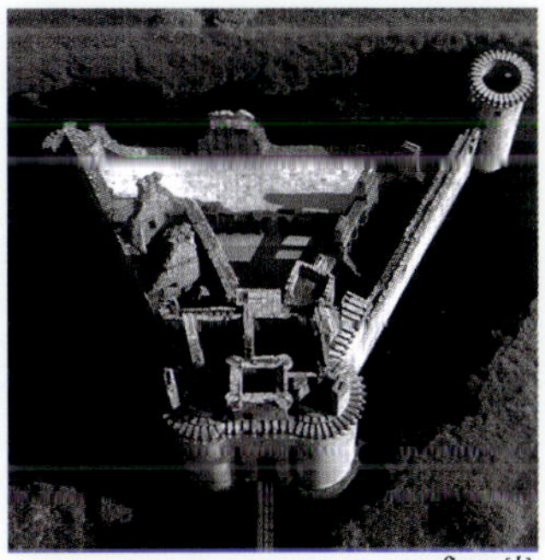

figure [b]

figure [a]

[*a*] & [*b*] Blast protection must address bombs of various sizes delivered by land or air, and the resultant impact and fire. The design must be such that a blast is unlikely to do damage due to the inherent strength of the structure. The curve of Timeship's circular outer wall provides an arch-like strength. The corners of the building are reminiscent of the towers of Medieval and Renaissance fortifications, and symbolize security as well as house lookouts.

[*c*] Timeship must be prepared for possible political, social, and other unrest in future decades.

figure [b]

TIMESHIP'S SECURITY PROGRAM was developed in a six-step process: First, a thorough understanding of the institution's mission and operation was developed and articulated. Next, the threats that could compromise Timeship in the immediate, mid-term, and long-term future were identified and analyzed based on historical knowledge, contemporary thinking, and future projections. Then rational and justifiable strategies to mitigate these threats were developed. Once this was accomplished, a partnership was set up between the security consultants and the design team to insure that security concerns were addressed at every stage of the design. Then every aspect of the design was evaluated against contemporary potential threats and future anticipated threats. After security was incorporated throughout the design, work began on the electronic aspects of the security system to monitor, control, detect, and deter threats.

figure [c]

figure [*a*]

LEGEND

a THIS ZONE PROVIDES A SECURITY "LOOK-OUT" FOR THE OUTSIDE PERIMETER OF THE BUILDING.

b PRIMARY ENTRY POINT RECEPTION FOR PERSONNEL AND VISITORS. A SECURITY CHECK IS PERFORMED PRIOR TO ENTERING A SECURE UNDERGROUND TUNNEL FOR PASSAGE INTO THE TIMESHIP BUILDING.

c SECONDARY ENTRY POINT AT FACILITY FOR SUPPLIES AND PATIENT ARRIVAL. ALL TRUCKS ARE UNLOADED, SECURED AND THEN PUT ON A TIMESHIP VEHICLE FOR ENTERING THE BUILDING.

d A SERIES OF STEPPED EARTH FILLED BARRIERS OF VARIES HEIGHTS TO PREVENT VEHICLE AND ON-FOOT ACCESS TO THE PERIMETER OF BUILDING.

e FORTIFICATION "RING-WALL" PROVIDES VISUAL STRENGTH AND PERMANENCE FOR THE BUILDING IMAGE. THE WALL'S CLADDING WILL ALLOW ADAPTION FOR FUTURE RESKINNING OR CLADDING TO MEET FUTURE THREATS NOT YET CONCEIVED.

f THE BASTION / OBSERVATION TOWERS PROVIDE A SECURITY PLACEMENT FOR ALLOWING A LINE-OF-SIGHT UNOBSTRUCTED VIEWS FOR BOTH THE EXTERIOR PERIMETER WALL AND INSIDE TOP LEVEL OF THE BUILDING.

g INTERNAL STANDOFF BUFFER ZONE PROVIDES PROTECTION FOR THE CRITICAL LOWER LEVEL - COLD STORAGE COMMUNITIES.

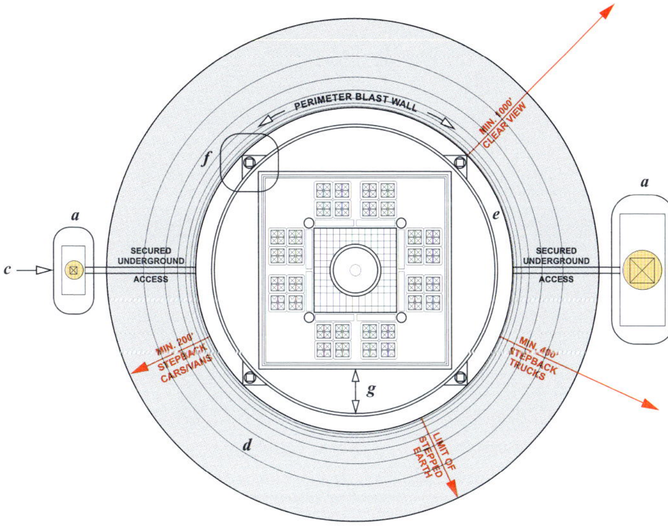

Building Site Plan Showing Security and Blast Design

figure [*b*]

TIMESHIP IS NOT UNIQUE in providing long-term secure storage, and the architect looked to facilities besides castles and bunkers for inspiration. For example, many insurance companies as well as other corporations and government agencies have obligations that require keeping records for over one hundred years. During the Cold War, NORAD needed its tracking and war fighting capabilities kept safe in the event of nuclear attack, and located facilities deep in mountain rock. However, it was decided that an aboveground design would be more appropriate, both physically and symbolically, for Timeship.

[*a*] NORAD, located in Cheyenne Mountain, housed several of the United States' key war fighting command centers. Its secure design inspired Timeship.

[*b*] A key element in security is standoff. The concentric rings surrounding Timeship will prevent vehicles from approaching too closely, thus providing protection from truck bombs and other potential threats.

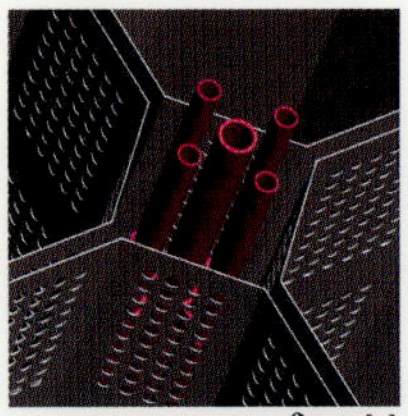

figure [*a*]

[*a*] Timeship components made of state-of-the-art materials to assure maximum performance and reliability.

[*b*] Three-dimensional Si composite nanostructures forming "nano cups" produced using chemical vapor deposition. Timeship is designed to permit retrofitting with new high-tech materials as needed.

[*c*] Model opened to show patient storage areas that will use evacuated stainless steel-walled chambers and superinsulation to assure patient temperature stability.

figure [*b*]

figure [*c*]

SINCE BUILDINGS BEGIN to deteriorate even before they are completed, and since Timeship must operate with high dependability for hundreds of years to assure the security of its passengers, it will be designed and constructed to take advantage of the most reliable advanced materials. Just as important, Timeship will be easy to maintain, with all building systems visible for regular inspection and designed so that any problem components can be identified and easily removed and replaced. We are now in a period of rapid development in building technology, with materials from aerospace such as composite carbon fiber and aerogels entering building construction, and the promise of building components made from carbon nanotubes in the future. Timeship will be constructed in such a way that components made from these new materials can easily be retrofitted into the building as they become available.

figure [*a*]

[*a*] Timeship model detail showing an airlock separating a transport service tube from a patient transfer volume. The security of Timeship's passengers and safety of its personnel depend on the reliability of its mechanical systems. Fire suppression will be handled by the abundant supply of LN_2, which can be used to extinguish any fire.

[*b*] Most buildings hide their mechanical systems. However, the Pompidou Center in Paris displays the mechanical systems on the outside as part of the building's aesthetic. Timeship's mechanical systems will also be poetically expressed.

figure [*b*]

WE USUALLY THINK OF ARCHITECTURE in terms of design, function, and structure, but in many buildings, especially in laboratories, the mechanical systems take up more than half of the space. Inspired by the Salk Center by Louis Kahn and the Pompidou Center by Renzo Piano and Richard Rogers, Timeship will celebrate its mechanical systems and make them an important part of its design, addressing the mechanical equipment with the same poetry and precision as the rest of the project. Mechanical systems in conventional buildings generally have a life span of twenty to twenty-five years. Timeship requires a rethinking of this approach and the use of highly durable mechanical equipment made with new corrosion-resistant materials, including ceramics, designed with redundancy and arranged to be easily replaced. The innovative patient cold storage area at Timeship involves few moving parts and is therefore highly robust.

[*a*] Timeship, like all buildings, is energy dependent, but it is designed to be in harmony with the ecosystem, using renewable energy sources where possible.

[*b*] The Japanese "Zen Garden" provides an example of a human relationship with nature in which nature is no longer wild, but where humans must be stewards of nature and live in harmony with it.

[*c*] Frank Lloyd Wright's winter home and studio, Taliesin West, relates to its Arizona landscape ecologically, aesthetically, and culturally, and as such provides an inspiration for Timeship.

figure [*a*]

figure [*b*]

TIMESHIP'S REANIMATED PASSENGERS will want to return to a future earth that is healthy, vital, and ecologically sound. Buildings, not automobiles, are now recognized to be the largest contributors to the greenhouse gasses that may be causing disruptive global warming. Therefore, in recognition of our responsibility for stewardship of the planet, Timeship's construction, operation, maintenance, and energy systems will all incorporate efficient, state-of-the-art, environmentally conscious technologies that will not only minimize Timeship's own dependency on carbon-based fuels, but will also serve as a model for future architecture.

Inspired by Frank Lloyd Wright's organic architecture, Stephen Valentine has designed Timeship to exist in harmony with the environment, relate to the ecology of its site, reflect the cultural traditions of its region, and enhance its spectacular setting.

figure [*c*]

figure [a]

[*a*] Below Prometheus Plaza, at the heart of Timeship, is the power plant that drives the entire complex and keeps the giant storage spheres filled with LN_2 to provide emergency backup energy.

[*b*] Turbines will generate electricity at Timeship. However, machines with moving parts are vulnerable to wear and breakdown, and future technologies may hold the promise of performing such functions with fewer moving parts and thereby increasing both efficiency and reliability. Eventually, Timeship may be able to generate electricity with such possible future green technologies.

figure [b]

TIMESHIP WILL BE HIGHLY EFFICIENT in its energy consumption and will use alternative renewable energy as well as be on the grid with two independent sources for reliability. Timeship's energy efficiency begins with super insulation in its patient cold storage areas, and extends to its laboratories and support spaces that are designed to recycle waste heat.

Timeship's onsite energy sources may include geomagmatic devices to generate electricity from ground heat with minimal moving parts, and advanced photovoltaic thin films to cover all appropriate surfaces and generate electricity from sunlight.

Because of the vulnerability of its cryopreserved patients, Timeship cannot permit any interruptions in energy availability for cooling, so it will store emergency backup energy onsite in the form of LN_2 to allow it to keep functioning for months in case of a disruption of outside sources.

06 SITE SELECTION

SITE SELECTION

THE SITE SELECTION PROCESS for Timeship was vital for several reasons. The building must be in a location that is safe from both natural hazards and human threats for up to hundreds of years, as well as have strong infrastructural support. This includes being near a major metropolitan area (at a distance of approximately fifty miles to provide protection from possible nuclear attack), and being near good roads and two international airports to provide reliable access for patients and for organs for transplantation. Timeship must also be located near research universities and other biotech facilities and in an area that provides a good quality of life for its scientists and technicians and their families. Finally, the site must accommodate not only the six-acre Timeship building and its immediate services, but also future possible development, including additional research laboratories, a conference center, housing, and a hotel.

In order to meet these requirements, Timeship architect Stephen Valentine, working with a team of consultants and a leading site selection firm, conducted one of the most exhaustive site selection studies ever undertaken, analyzing dozens of criteria for all of the contiguous forty-eight states of the United States.

In the first stage of this process, a ten-kilometer grid was applied to the United States and each of the squares of the grid was analyzed with a broad series of screening criteria. Some of these criteria were "fatal flaws," such as earthquake zones that absolutely eliminated an area even if it was otherwise favorable.

Over the past several decades, intensive research has led to building codes and engineering approaches that make buildings safer in areas of intense earthquake activity. However, the general principle of such design approaches is that people should be able to get out of the building safely and the building should remain standing, even if it might need extensive repairs or be demolished. In the case of a hospital, patients might be evacuated and moved to another hospital. These standards are not applicable to Timeship, since Timeship's patients and equipment are fragile and the patients should not be removed until they are reanimated. Earthquakes are unpredictable. While historical information can suggest areas of greater risk, there is no guarantee that one won't occur even in an area of little activity. With the need for the security of Timeship's contents, a site with very low seismic risk was sought. At the same time, the building will be designed with a high degree of earthquake resistance, both in its structural and mechanical equipment.

After areas with fatal flaws were eliminated, a Multi-Attribute Utility Function Analysis (MAUFA) was applied to the remaining areas. The MAUFA is a proprietary algorithmic procedure in which attributes (or objectives, indicators, or criteria) are listed and each attribute is weighed against the others, thus expressing value judgments numerically (such as it is good to be near an airport, or bad to be near a nuclear reactor). Then, for a given site, each one of the many attributes is considered and given a numerical evaluation. The numbers for the attribute's weight and the evaluation are then combined using the algorithmic procedure to yield a final numeric score. Favorable and unfavorable scores can then be compared. The list of attributes to be evaluated was developed through extensive studies by the site selection team and outside consultants. The complex results of this process are then represented on computer-generated maps. When this approach is applied against two or more potential sites, their scores can be objectively compared. Of course, subjective criteria also enter the picture, but this method enables decision makers to be aware of when they are being objective or subjective.

Several areas were chosen within the regions found to be more favorable and analyzed at a finer scale using a two-kilometer grid, and more refined criteria. For example, in this second stage, economic development incentives were taken into account. Which states and regions want this kind of biotech development? Will they provide tax breaks or otherwise help economically? The result of the second stage was a specific indication of favorable areas. At this stage, several sites were again evaluated with the MAUFA. In the third and final stage, the sites under consideration were evaluated once more to be sure all issues were addressed.

Security was a primary concern in the site selection process, since Timeship will store precious biological materials and cryopreserved patients, and must keep these contents safe for up to hundreds of years. Therefore the location must be as safe as possible from earthquakes and volcanoes; hurricanes, tornadoes, blizzards and other storms; floods; the possibility of rising sea levels due to global warming; domestic and global terrorism, military action, and threats at the borders; and interruptions of energy supplies.

For the most part, even severe storms would not be a threat to a substantial building like Timeship, but they could cause interruptions to the power supplies, so there will be highly redundant auxiliary power. In addition, storms of all kinds are a serious consideration in that they might inhibit access to the site for new patients while they are in time-sensitive conditions, for example cooled down but not yet vitrified, or in a transport pod with only limited coolant. For this reason, location in a region subject to snowstorms that might close down airports for several days was rejected. Flooding is also a serious consideration, as water exposed to liquid nitrogen would rapidly freeze and could cause structural damage. Therefore, low-lying areas were also rejected.

The site selection for Timeship also needed to take into consideration a range of potential human threats. Some version of the Cold War might reignite, bringing an increased threat of nuclear attack against the United States. There might be an attack over a border. Domestic or foreign terrorists might attack with conventional or nuclear weapons. There might be an upheaval of social unrest. A nuclear power plant might suffer an accident. And there might be any of several forms of threats specifically against Timeship for various reasons.

One approach to these potential threats is to design a highly secure building, which is addressed elsewhere in this book. But the best way to address them is to locate Timeship where they are least likely to occur, and where local citizens are likely to be supportive of Timeship's mission. Toward this end, Valentine and his team created a set of criteria establishing minimum distances of Timeship from borders, nuclear power plants, and from military installations and population centers of over 100,000, both of which are considered potential targets for nuclear attack.

While the site for Timeship must be remote from potential hazards, it must also have strong positive features. It must have good transportation access in order to bring in new patients and to allow timely delivery of organs for transplant. It should be near a major research university and other biotech facilities to assure support and a pool of qualified high tech workers. And Timeship should be located where there is a good quality of life. Research scientists and highly skilled workers are in demand and have their choice of jobs. Major factors in their choice are often good housing, reasonable cost of living, and the cultural, recreational, and educational opportunities available to them and their families. Timeship's location must offer a rich combination of all of these.

Finally, there is the issue that thousands of people will be spending up to hundreds of years at Timeship before they might be reanimated. While it might seem that these people would not care about the aesthetics of their environment while they are suspended, in fact many potential clients have stated that they do care. For this reason, Valentine has chosen the landscape of Tuscany, revered for its classical heritage and its beauty, as an inspiration for the choice of the Timeship site.

TIMESHIP WILL BE the temporary waiting place for thousands of cryopreserved clients traveling to the future, as well as the workplace for some of the leading biotech researchers. Besides the building itself, Timeship's site will eventually include additional laboratories, support facilities, and a small conference center. As one of his inspirations in selecting a site, the architect chose the landscape of Tuscany, the birthplace of the Renaissance: its classical beauty matches the timelessness of Timeship's architecture. The building will be nestled securely among rolling hills that provide both an appealing setting and natural protection.

figure [*a*]

figure [*b*]

[*a*] & [*b*] Educational, recreational, and cultural availabilities, along with satellite, cable, and the Internet now make even remote areas livable. Climate and physical beauty become important in choosing a location, and Tuscany serves as an inspiration.

[*c*] The Timeship site will be approximately 1,000 acres that will be developed with related support facilities.

figure [*c*]

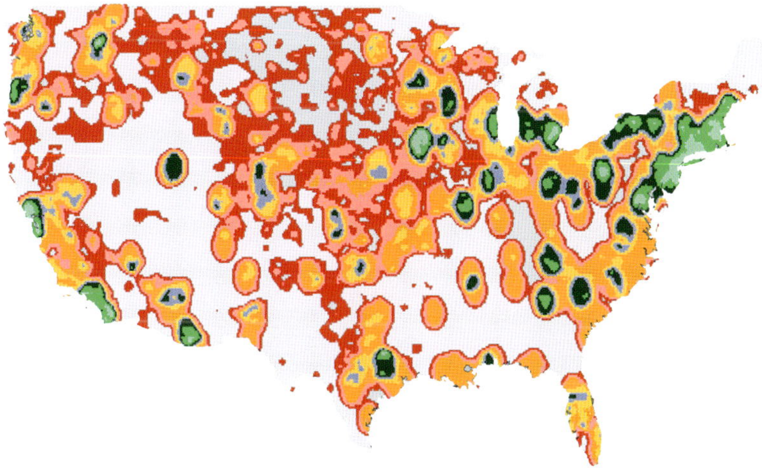

figure [*a*]

figure [*b*]

TIMESHIP MUST BE LOCATED about fifty miles from a major metropolitan area and near research universities and other biotech companies to assure adequate infrastructure, easy access to support facilities, and availability of a qualified staff. It also must be near good roads and, most importantly, two international airports to provide quick access for patients to be cryopreserved and organs retrieved for transplantation.

Timeship should be located in a place where scientists and technicians will want to live with their families. Modern telecommunications and the spread of quality retail chains have made just about every part of the country livable for workers and functional for industry. Companies can now choose where to locate based on quality of life as defined by attractive housing, good schools, recreation, cultural facilities, low crime, reasonable cost of living, and, of course, good climate and a beautiful setting.

As vitrification is perfected and more people choose to become cryopreserved, a broad base of support for cryonics will be needed. Timeship will play a major role in generating that support. Its setting, as well as the design of the building, will help it fulfill that role.

[*a*] The site analysis for Timeship began with things to be avoided, and then moved to things that are desired, including good quality of life, good infrastructure including transportation, and close proximity to a major research university and biotech industry, as shown on this map (altered for confidentiality).

[*b*] An inspirational setting is important for the Timeship site.

SINCE TIMESHIP WILL MAINTAIN cryopreserved biological material, including human patients, it must be in a location safe from natural threats for up to hundreds of years. The most important consideration is earthquakes, which can damage the fragile patients and sensitive control systems. Also considered were storms that might not threaten the building, but could disrupt power or quick access to Timeship for new patients in a time-sensitive condition. To assure the long-term safety of Timeship, the architect and his consulting site selection team analyzed these and dozens of other criteria, including glacial movements and rising sea levels due to global warming, as well as volcanic eruptions and meteor strikes. Although most of the United States is safe and habitable for the majority of uses, the requirements for Timeship are far more stringent.

figure [*a*]

figure [*b*]

[*a*] Earthquakes that cause moderate damage to other facilities would be devastating to Timeship due to sensitive equipment and patients who cannot be evacuated.

[*b*] Severe snow storms could delay arrival of patients and also the shipment of organs for transplantation.

[*c*] Powerful hurricanes could cause damage and disrupt access to Timeship for days.

figure [*c*]

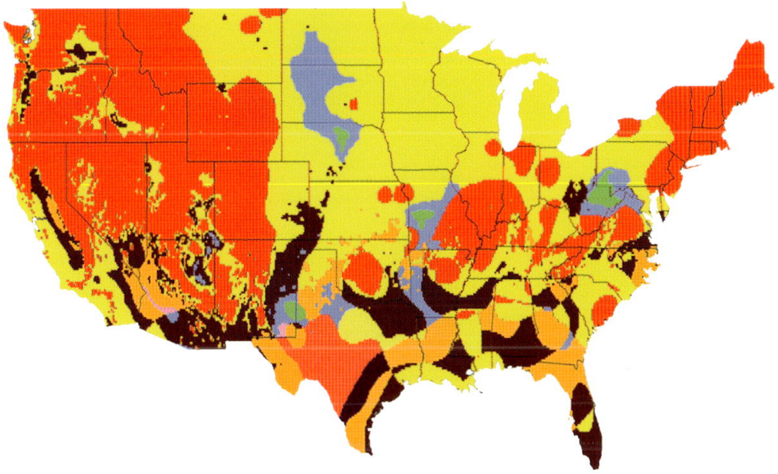

figure [*a*]

figure [*b*]

BECAUSE OF THE IMPORTANCE of locating Timeship in an area of least natural threat, the design team developed unique site selection methodologies, including the Multi-Attribute Utility Function Analysis (MAUFA). First, a ten-kilometer grid was placed over the contiguous United States. (It had been decided for various reasons to locate Timeship in one of the contiguous forty-eight states.) Then, all of the likely potential natural threats were identified, including earthquakes, storms, and floods, as well as less likely threats, such as volcanoes and extreme changes in sea level. A complex series of values were then assigned to each of these threats, and a proprietary computer algorithm was developed to compile the values for all of the threats in each of the squares of the grid. The result was a series of maps, dense with information, that showed more favorable and less favorable potential locations for Timeship. This information was later combined with an analysis of human threats, as well as favorable conditions such as infrastructural support, other nearby biotech facilities, and quality of life to make the final determination of acceptable potential locations. This determination then became the basis of more fine-grained local searches to find the actual site.

[*a*] One of the many maps generated by the detailed analysis for the location of Timeship. The patterns generated by the algorithms give a quick indication of which are acceptable and unacceptable regions. (Map altered for confidentiality.)

[*b*] Tornadoes and storms could delay the delivery of patients to Timeship while they are in a fragile and time-sensitive condition.

figure [a]

AS WITH NATURAL HAZARDS, Timeship must also be concerned about human hazards and threats. The best defense against such hazards and threats is to be located where they are least likely to occur. Of primary concern are accidents at nuclear power plants, which Chernobyl showed can be far-reaching; nuclear attack, which might come from terrorists as well as other nations; and attacks over U.S. borders, which appear more porous every day. Other concerns included locations near major population centers, highways, airports, and military bases. Also considered are rises in sea level due to global warming that might put low-lying areas under water at some future time. To address these concerns, a series of detailed surveys were done paralleling those done for natural hazards.

figure [b]

[*a*] Global warming may lead to a rise in sea levels. The Timeship site will be situated at a safe elevation.

[*b*] Hiroshima. Timeship will be located at least 50 miles from military bases and cities of 100,000 or more to reduce danger from nuclear attack.

[*c*] A close-scale mapping of potential human threats was done for each area under consideration.

figure [c]

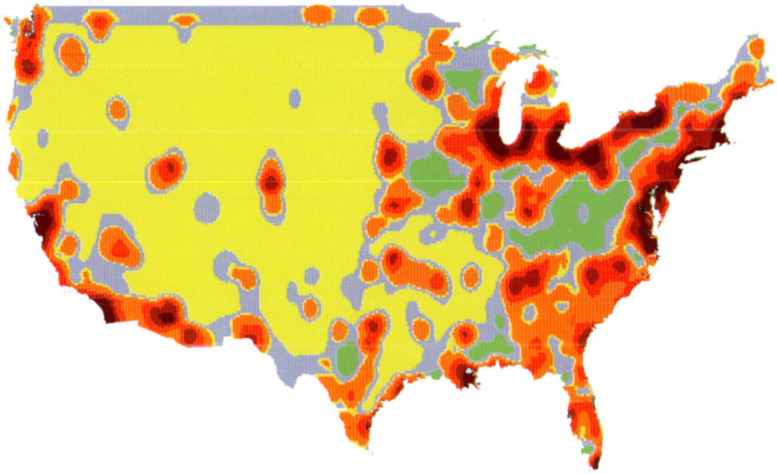

figure [*a*]

figure [*b*]

[*a*] Map generated from a security algorithm applied to a 10-kilometer grid to identify potential locations that have minimal exposure to human hazards and threats. (Map altered for confidentiality.)

[*b*] Oklahoma bombing at the Murrah Federal Building. Domestic and foreign terrorist attacks can pose a serious threat to Timeship, so its location must be carefully considered.

THE UNITED STATES HAS, for the past two decades, experienced a high degree of social and economic prosperity and stability. But when one thinks of two centuries, such upheavals as the Civil War come into view, and as recently as the past forty years the country experienced riots, looting and burning, and political assassinations. More recently, we have experienced major terrorist attacks of both domestic and foreign origin.

In addition, despite the end of the Cold War, the threat of nuclear attacks remains a concern for Timeship, both in terms of its design and its location. In a time frame of one hundred years or more, the United States could fall into a confrontational relationship with a nuclear power, or a rogue state or terrorist group could launch or smuggle nuclear weapons into the country. For this reason Timeship will be located approximately fifty miles from potential targets. In addition, Timeship will be located from twenty to one hundred miles from borders and coasts.

Timeship will be operational for possibly centuries and has to anticipate both known and as yet unknown threats. Timeship's best defense against such threats is a highly secure design and location in a place where they are least likely to occur.

figure [a]

THE SITE SELECTION began with applying a series of broad criteria to large areas. It was eventually scaled down to the specific features of the terrain that should surround Timeship. These features included seclusion; good views, perhaps of hills in the distance; a raised spot on which to place the building; and the possibility of surrounding Timeship with a moat-like water feature.

Once several potential sites were selected, each was digitized and analyzed in detail through computer modeling to determine how Timeship would fit. The computer models allowed for different positioning of the building and for the study of views to and from the building.

figure [b]

[*a*], [*b*], & [*c*] Different views of a site under study. A computer-generated model of Timeship was placed in a digitized model of the site, which then permitted positioning the building in various locations to study each from several different points of view.

Opposite page: An arial view of a model of Timeship showing the surrounding stepped rings.

figure [c]

GLOSSARY

ANTI-AGING: Refers to a broad range of approaches, from nutrition to genetic engineering, to extend the healthy life span or to stop aging altogether.

BIOSTASIS: Complete arrest of biological activity. *Cryopreservation* is a form of biostasis.

CELL REPAIR DEVICE: A theorized future nanomachine combining sophisticated sensory, manipulative, and data processing functions on a molecular scale, and programmable for biological repair. An example of *nanomedicine*.

COMMUNITY: A unit of Timeship. People and organs are placed in *pods*. Hundreds of *whole-bodies* and thousands of organs can be held in pods in *Neighborhoods*. Neighborhoods are grouped in three-by-three grids of nine to form communities. There will be twelve communities at Timeship.

CRYO: Prefix referring to extreme cold.

CRYOBIOLOGY: The branch of biology that studies life at below-normal temperatures. Usually cryobiology deals with the effects of freezing and thawing. However, any temperature below what is normal for any given living system falls into the realm of cryobiology, including hypothermia, hibernation, natural frost hardiness of insects and plants, and medical organ preservation at low temperatures.

CRYOGENICS: The branch of physics that studies the causes and effects of extremely low temperatures. Cryogenics usually relates to cryotechnologies such as producing liquefied gasses as well as to many other low-temperature physical effects such as superfluidity and the behavior of Bose-Einstein condensates.

CRYONIC SUSPENSION: The practice of keeping a legally dead *cryopreserved* (see *cryopreservation*) human body or brain at an extremely low temperature in the hope of eventual *reanimation* with the help of future medical technologies.

CRYONICS: See *cryonic suspension* and *cryopreservation*, which are the terms now most commonly used.

CRYOPREPARATION: Preparing a patient for *cryopreservation*. The process is as complex as modern surgery, and includes infusion with *cryoprotective agents*.

CRYOPRESERVATION: The process of preserving and storing living systems in a viable condition at low temperatures for future use. Traditionally, cryopreservation has meant preservation by freezing, and the word is still used with this meaning in many cases. However, the term can also cover preservation by *vitrification*, or ice-free cryopreservation. The terms *cryonics* and *cryonic suspension* are usually used when referring to people, while "cryopreservation" can refer to people, organs, or tissue.

CRYOPRESERVED: See *cryopreservation*.

CRYOPROTECTANT: Alternate term for *cryoprotective agent*.

CRYOPROTECTIVE AGENT: A chemical inhibiting or reducing the formation of damaging ice crystals in biological tissues during cooling. Glycerol and DMSO are examples. *Cryoprotectants* are used in *cryonic suspension* procedures to minimize such damage.

CRYOSTORAGE: The storage in highly controlled conditions of a *cryopreserved* patient or biological material.

DEATH: Definitions of death have changed over time. For the approach taken in this book, death is the irreversible loss of the structural information that encodes memory and personality. By this definition, if *cryopreserved* patients are eventually *reanimated*, they were not dead.

DEATH, BRAIN: Condition making restoration of normal brain function impossible through present medical technology. It is now known that the initial stages of brain death are sometimes not inherently irreversible. Later stages may be reversible by future technologies.

DEATH, CLINICAL: Absence of respiration and heartbeat. This was once thought to constitute irreversible loss of life, but advancing medical technologies have sometimes made the early stages of clinical death a reversible condition. Medicine has only begun to overturn historic death criteria, and even prolonged clinical death may be one of a class of conditions future medicine will regard as treatable.

DEATH, INFORMATION-THEORETIC: Damage to the human brain and the information in it so severe that a person's memories, personality, hopes, dreams, etc. have been destroyed beyond the possibility of recovery by any future technology. The concept of information-theoretic death was introduced in response to advances in medical technology that have continued to make previous definitions of death obsolete. We currently do not know enough to be able to state when information-theoretic death occurs.

DEATH, LEGAL: Irreversible loss of life for legal purposes. Criteria for legal death vary from jurisdiction to jurisdiction and often do not reflect resuscitation capabilities of even today's medicine. In actual medical practice, legal death does not always mean that the basis of life has been lost, only that a qualified authority sees no value in continuing to support life with technology. Declaration of legal death is currently a prerequisite for *cryonic suspension* procedures.

DEWAR, CRYOGENIC: Super-insulated vacuum-jacketed storage vessel used for holding liquid nitrogen (LN_2) and *cryonic suspension* patients. Dewars containing LN_2 are also commonly used for all kinds of *cryostorage*.

EXTREME LIFE EXTENSION: The term *life extension* refers to any extension of the healthy life span. "Extreme life extension" refers to life spans that might extend into the hundreds of years and to a state in which death comes only from trauma. (See *Practical immortality*.)

GENETIC ENGINEERING: The alteration of DNA to produce novel traits. Proponents of *extreme life extension* propose genetic engineering as one means of eliminating aging and otherwise improving human beings.

LIFE EXTENSION: An increase in the life span. For those who regard aging as a disease, therapeutic methods to achieve maximum life span are *anti-aging* medicine.

LN_2: Abbreviation for liquid nitrogen, the agent presently used for maintaining *cryopreserved* patients at a constant temperature of –196°C. At this temperature, biological tissue can be maintained unchanged indefinitely. Patients at Timeship will be kept at temperatures higher than –196°C, and therefore will not be stored in LN_2. However, LN_2 will be used at Timeship to produce cooling through evaporation.

MANDALA: A symbolic diagram, either imagined or depicted, typically consisting of concentric circles and squares, with a central symbol that can be a figure. In practice, mandala has become a generic term for any concentric plan, chart, or geometric pattern that represents the cosmos metaphysically or symbolically.

NANO: Prefix referring to sizes pertinent to *nanotechnology*. A nano-meter is one billionth of a meter.

NANOROBOT: Term derived from *nanotechnology* and robot, sometimes referred to as a nanobot. A hypothetical very small robot. In the medical field, such robots would be capable of swarming through a patient's body and effecting repairs of cells and even of DNA.

NANOMEDICINE: A medical field that envisions tiny devices such as *nanorobots* that will enter a patient's systems and engage in diagnosis, treatment, and repairs.

NANOTECHNOLOGY: Engineering at the molecular level. As proposed by Eric Drexler, nanotechnology includes molecular manufacturing: nanomachines assembling atoms and molecules to make anything economically. The medical implications of nanotechnology are central to many ideas in this book. Nanotechnology is advancing rapidly in numerous fields.

NEIGHBORHOOD: A unit of Timeship. (See *Community*.)

NEURO: *Patients* who choose to have only their brains *cryopreserved*. This choice is made based on the belief that the brain is the seat of memories and "self," and that future technologies will be able to create some kind of body around their brain when they are reanimated.

NEURO-SUSPENSION: A *cryopreservation* option in which only the brain is preserved. (See *Neuro*.)

PATIENT: The preferred term for a *cryopreserved* person, or a person undergoing the process of *cryopreservation*.

POD: A single unit in which a *cryopreserved* patient or organ is placed for *cryostorage*. The pod must provide protection, insulation, and uniform distribution of cold, and must fit in the cold storage area of Timeship with other pods.

PRACTICAL IMMORTALITY: *Extreme life extension* in which the only cause of death is trauma such as from an accident. The word "practical" is added to avoid confusion with metaphysical immortality in which there is no death, a condition perhaps exclusive to the gods.

REANIMATION: The future revival of *cryopreserved* patients. At this time reanimation of even a mouse has not been achieved, but human cells, tissues, and embryos are routinely frozen and revived, and there is reason to believe that scientific advances will eventually make reanimation possible.

SACRED GEOMETRY: Forms, patterns, proportions, intervals, and relationships that are found in many cultures and exist in our bodies, in our perceptions, in living creatures, in human creations, that reflect the structures of the universe.

SUSPENDED ANIMATION: State of greatly slowed or stopped metabolism from which organisms can reliably be restored to healthy function. Though long featured in science fiction, true suspended animation technology does not yet exist for humans, and this concept should not be confused with *cryopreservation*. There is presently no means of recovering patients placed in *cryopreservation* and will not be for a long time. Only when viewed from the broadest historical context—one considering the future existence of advanced cell repair technology—does *cryopreservation* possibly constitute a form of suspended animation.

TEMPERATURE CONTROLLED VOLUME (TCV): *Cryopreservation* storage container with Timeship features. *Patients* may be placed in TCVs before Timeship is operational.

TRANSHUMAN: Possible future entities that are no longer human. Rapid developments in biotech and the merging of electronics with humans suggest the possibility of changes in human nature, or even the possibility of moving beyond what we now understand as human. Transhumanism is a discipline that addresses this area.

TRANSPORT CONTAINER: A container in which a *pod* containing a *cryopreserved patient* will be placed for transport to Timeship or a TCV. The transport container will provide sufficient insulation as well as impact protection for the duration of the trip.

VITRIFICATION: In *cryobiology*, preservation at extremely low temperatures without freezing. Freezing involves ice crystal formation, which damages delicate structures such as blood vessels. Vitrification instead involves the formation of a glassy or amorphous solid state without crystals, and which, unlike freezing, is not intrinsically damaging to complicated living systems.

WHOLE-BODY SUSPENSION: A *cryopreservation* option in which a patient's entire body is preserved, as opposed to just the brain.

IMAGE CREDITS

CHAPTER ONE: ORIGINS OF TIMESHIP
Page 17
EARLY CONCEPT SKETCH-1, Stephen Valentine, Architect
Page 18
EARLY CONCEPT SKETCH-2, Stephen Valentine, Architect
Page 20
[a] BEN FRANKLIN: SESQUICENTENNIAL, © Norman Rockwell Family, [b] PROSPECT OF IMMORTALITY: BY ROBERT ETTINGER, © Robert Ettinger, [c] SLEEPING BEAUTY, © Disney Enterprises, Inc.
Page 21
[a] SLEEPER: DVD COVER, Courtesy of MGM Clip+Sill, [b] A.I. ARTIFICIAL INTELLIGENCE, © 2001 Warner Bros., Time Warner Entertainment Co. and DreamWorks LLC., [c] 2001: A SPACE ODYSSEY © Turner Entertainment Co. A Warner Bros. Enterainment Co.
Page 22
[a] MUMMY EXCAVATION, Aladin Abdel Naby/Corbis, [b] HEAD ON ROBOT, © Street and Smith Publications, [c] DR. FRANKENSTEIN, © Universal Pictures
Page 23
[a] HUMAN HEART TRANSPLANT, © M. Peterson/Redux, [b] EMBRYO: 16 CELL STAGE, Dr. Yorgos Nikas/PR Inc., [c] DR. BEDFORD, FIRST CRYONICS PATIENT, Alcor Life Extension Foundation
Page 24
[a] PATIENT CRYOPRESERVATION, [b] CRYOPRO-TECTANT SUPPLY, Photos by Charles Platt, courtesy of Alcor Life Extension Foundation
Page 25
[a] NEURO CRYO CONTAINER, [b] PATIENT DEWAR: TOP COVER OPENED, Photos by Charles Platt, courtesy of Alcor Life Extension Foundation, [c] DEWARS: PATIENT STORAGE, Courtesy of Alcor Life Extension Foundation
Page 26
[a] TIMESHIP PROGRAM WORKSHOP TEAM, [b] TIME-SHIP PROGRAM REPORT, [c] TIMESHIP WORKSHOP NOTES, Stephen Valentine, Architect
Page 27
[a] TIMESHIP MAS EXHIBIT, Photo by Piotr Redlinski, Stephen Valentine, Architect, [b] CONCEPT FLOOR PLANS, Stephen Valentine, Architect

CHAPTER TWO: LIFE EXTENSION RESEARCH
Page 29
DNA MODEL: PLAN VIEW, ALFRED PASIEKA/PR Inc.
Page 30
DNA MODEL, Peter Gregorie/Photos To Go
Page 32
[a] FREEZING VS VITRIFICATION, Stephen Valentine, Architect, [b] GLYCEROL-PROTECTED DIAGRAM, [c] VITRIFIED-PROTECTED DIAGRAM, [d] PRO-POSED WATER BONDING MOLECULES, © 21st CM
Page 33
[a] VITRIFIED VS FROZEN KIDNEYS, © G.M. Fahy, [b] SUPERCOOL X-1000 ICE BLOCKER, © 21st CM
Page 34
[a] HUMAN BRAIN, Mehau Kulyk/PR Inc., [b] ELECTRON MICROGRAPH: BRAIN TISSUE, © 21st CM, [c] VITRIFIED BRAIN TISSUE, © 21st CM
Page 35
[a] MAMMALIAN HIPPOCAMPAL SLICE, © 21st CM, [b] ICE CRYSTALS, © Alaska Stock.com
Page 36
[a] NANOROBOT: MEMORY DEVICE, Stephen Valentine/J. Anderson, [b] NANOROBOT: CELL REPAIR DEVICE, [c] NANOROBOT: CELL REPAIR DEVICE DRAWING, Yuriy Svidinenko
Page 37
[a] NANOROBOT: INJECTOR DEVICE, C. Jay Conely/PR Inc., [b] NANO MOLECULAR MANUFAC-TURING, George Canciani, Foresight Nanotech Institute, [c] NANOROBOT HUNTER UNIT, Rendering by Jeff Johnson, Hybrid Medical Animation
Page 38
[a] DNA SEQUENCING, Jacob Halaska/Photos To Go, [b] KAMATO HONGO: AGE 116, JIJI Press/AFP/Getty Images, [c] BOB THE DOG, Courtesy of Critical Care Research
Page 39
[a] ARTIFICIAL WOMB, Fritz Goro/Getty Images, [b] YEL-LOW SEAHORSE, Photo by Lisa Young, [c] PHYSICIAN REVIEWING BRAIN SCAN, © Pete Saloutos/Corbis

CHAPTER THREE: ARCHITECTURE
Page 41
WATER DROPLET ON FILM, Unknown Source
Page 42
EARLY CONCEPT IMAGE-3, Stephen Valentine, Architect
Page 44
[a] DALI'S PERSISTENCE OF MEMORY, © 2005 Salvador Dali, Gala-Salvador Dali Foundation / Artists Rights Society (ARS), New York, [b] TREE CROSS-SECTION, Stephen Valentine, Architect, [c] DIVER BACK FLIP: STOP ACTION PHOTO, ©Harold & Esther Edgerton Foundation, 2006, Palm Press, Inc.
Page 45
[a] PYRAMID-SPHINX AT GIZA, © Corbis, [b] STONE-HENGE, © Stonehenge Images, [c] OBSERVATORY JAIPUR, Magnum-Rodgers
Page 46
[a] STAR TRAILS: TIME LAPSE PHOTO, Gemini Observatory / Association of Universities for Research in Astronomy, [b] TIMESHIP RINGS P10 EXPANSION, Stephen Valentine, Architect
Page 47
[a] DIATOM: AMPHITETRAS ANTEDILUVIANA, Perennou Nuridsany/PR Inc., [b] DIATOM: ACTINOP-TUCHUS HELIOPELTA, Eric Grove/PR Inc., [c] PLAN OF TIMESHIP, Stephen Valentine, Architect
Page 48
[a] MILK DROP CROWN, Clive Kocher/PR Inc., [b] TIMESHIP MODEL FROM ABOVE-1, Stephen Valentine, Architect, [c] WATER DROPLET SPLASH, Composite of photos by Andrew Davidhazy, Rochester Institute of Technology
Page 49
[a] MODEL: ARISTOTLE PLAZA, [b] TIMESHIP ICONO-GRAPHY: FLOWER-1, Stephen Valentine, Architect
Page 50
[a] DA VINCI MAN - MRI, Simon Frasier/PR Inc., [b] SQUARED CIRCLE DIAGRAM-1, [c] TIMESHIP PLAN DIAGRAM, Stephen Valentine, Architect
Page 51
[a] TAJ MAHAL: Mosque Ceiling, Stephen Valentine, Architect, [b] CACTUS, Photo by: Anne M.Archibald, [c] NAU-TILUS SHELL: GNOMONIC GROWTH, S. Sammons/PR Inc.
Page 52
[a] MANDALA: BINDU, Ajit Mookerjee collection, © Thames and Hudson, [b] MANDALA: SRI YANTRA HINDU, [c] MANDALA: MEXICAN AZTEC SQUARE, © Thames and Hudson
Page 53
[a] MANDALA: CHRISTIAN FRANCE 14TH CENTURY, © The British Museum, [b] MANDALA: TAOIST CHINESE, © Thames and Hudson, [c] MODEL: TIMESHIP PLAN, Stephen Valentine, Architect
Page 54
[a] MOSQUE OF IBN, Hulton Archieve/Getty Images, [b] CHARTRES CATHEDRAL ROSE WINDOW, Stephen Valentine, Architect, [c] CORDOBA MOSQUE CEILING, L. Castaneda/ Getty Images
Page 55
[a] MODEL: TIMESHIP OVERALL, Stephen Valentine, Architect, [b] DELPHI TEMPLE, Jose Fuste Raga/Corbis
Page 56
[a] PHOENIX PAINTING, University of Aberdeen, [b] PHOENIX TEMPLE, [c] ARISTOTLE PLAZA FROM ABOVE, Stephen Valentine, Architect
Page 57
[a] SATURN'S RINGS, NASA / JPL / Space Science Institute, [b] SOLAR SYSTEM, [c] TIMESHIP PLAN INDI-CATING RINGS-ORBITS, Stephen Valentine, Architect
Page 58
[a] TIMESHIP ELEVATION OUTER WALL DRAWING, [b] TIMESHIP MODEL FROM ABOVE-2, Stephen Valentine, Architect

Page 59
[a] TIMESHIP ICONOGRAPHY: FLOWER-2, Carter Harmon, [b] CONE AND WALL ELEVATION, Stephen Valentine, Architect, [c] FLOWER PISTIL AND STAMENS, C.Nuridsany & M.Perannou/Science Photo Library
Page 60
[a] ROMAN PANTHEON, Erich Lessing/Art Resource, [b] PROMETHEUS PLAZA LOOKING UPWARD, [c] MODEL: PROMETHEUS PLAZA, Stephen Valentine, Architect
Page 61
[a] ARISTOTLE RENDERED PLAN, [b] ARISTOTLE PLAZA CAD WIREFRAME, Stephen Valentine, Architect
Page 62
[a] TIMESHIP ELEVATION, [b] TIMESHIP SECTION, Stephen Valentine, Architect
Page 63
TIMESHIP SITE PLAN, Stephen Valentine, Architect
Pages 64–65
TIMESHIP AT SUNSET, Stephen Valentine, Architect

CHAPTER FOUR: CRYOPRESERVATION
Page 67
PLAN MODEL: CRYO STORAGE LEVEL, Stephen Valentine, Architect
Page 68
NEIGHBORHOOD WITH DA VINCI MAN, Stephen Valentine, Architect
Page 70
[a] 3D PLAN: COMMUNITY-1, [b] ACCESS PLAN: COLD STORAGE COMMUNITIES, [c] TIMESHIP STORAGE UNITS, Stephen Valentine, Architect
Page 71
TIMESHIP PLAN CRYO STORAGE LEVEL, Stephen Valentine, Architect
Page 72
[a] NEIGHBORHOOD MODEL, [b] 3D PLAN COMMUNITY-2, Stephen Valentine, Architect
Page 73
[a] DRAWING: NEIGHBORHOOD LAYOUT, [b] DRAWING: NEIGHBORHOOD EARLY SCHEMES, [c] DRAWING: FAMILY CLUSTER, Stephen Valentine, Architect
Page 74
[a] COMMUNITY MODEL: VIEW 1, [b] COMMUNITY MODEL: VIEW-2, [c] COMMUNITY MODEL: VIEW-3, Stephen Valentine, Architect
Page 75
[a] COMMUNITY AIRLOCK DOOR, Stephen Valentine, Architect, [b] NASA AIRLOCK, NASA
Page 76
[a] DRAWING: COMMUNITY LN_2 PLAN, [b] 3D PLAN: COMMUNITY TOP ENCLOSURE, Stephen Valentine, Architect
Page 77
[a] ELECTRIC SUPPLY DIAGRAM, [b] CRYO LN_2 DIAGRAM, [c] LN_2 TANKS MODEL, Stephen Valentine, Architect
Page 78
[a] COOL-DOWN DIAGRAM A, [b] COOL-DOWN DIAGRAM B, [c] PATIENT COOL-DOWN CAPSULE, Stephen Valentine, Architect
Page 79
[a] DRAWING SECTION THROUGH COMMUNITY, [b] 3D SECTION AT COMMUNITY, [c] 3D ELEVATION AT COMMUNITY, Stephen Valentine, Architect
Page 80
[a] MERCURY SPACE CAPSULE, NASA, [b] SARCOPHAGUS, SCALA/Art Resource, [c] FULL BODY POD AND SLEEVE, Stephen Valentine, Architect
Page 81
DRAWING PLAN TCV DEWAR, Stephen Valentine, Architect
Page 82
[a] SLIM FIGURE, Unknown Source, [b] OBESE FIGURE, "Dolly Dimples" c.1940, [c] DRAWING PLAN PATIENT POD, [d] DIAGRAM: MAN AT 99 PERCENT, Stephen Valentine, Architect
Page 83
[a] DRAWING: SECTION AT TRANSPORT CAPSULE, [b] DRAWING: ISOMETRIC AT TRANSPORT CAPSULE, [c] TRANSPORT CAPSULE WITH PATIENT, Stephen Valentine, Architect
Page 84
[a] TCV DEWAR PARTS, [b] DRAWING: PLAN TCV DEWAR, Stephen Valentine, Architect
Page 85
[a] SECTIONS THROUGH TCV DEWAR, [b] CONDUIT BRAIDS, [c] HEAT FLOW DIAGRAM-A, [d] HEAT FLOW DIAGRAM-B, Stephen Valentine, Architect
Page 86
[a] 3D WHOLE-BODY POD PARTS, [b] WHOLE-BODY BODY SLEEVE, Stephen Valentine, Architect
Page 87
[a] NEURO-ORGAN POD PARTS, [b] NEURO-ORGAN POD TOWER, [c] NEURO-ORGAN POD, [d] NEURO-ORGAN POD IMPACT DIAGRAM, Stephen Valentine, Architect

CHAPTER FIVE: BUILDING TECHNOLOGY
Page 89
INTERNATIONAL SPACE STATION, NASA
Page 90
CRYO-STORAGE ASSEMBLY, Stephen Valentine, Architect
Page 92
[a] HARLECH CASTLE, © Associated Press, [b] WW-II BUNKER, Stephen Valentine, Architect, [c] WW II BUNKER PLANS, Stephen Valentine, Architect
Page 93
[a] TIMESHIP MODEL ELEVATION VIEW, Stephen Valentine, Architect, [b] CAERLAVEROCK CASTLE, © Colin Baxter
Page 94
[a] MODEL: PLAN RAMPART-BASTION, [b] MODEL: CORNER BASTION, Stephen Valentine, Architect, [c] SOCIAL UNREST: AP Images/Alan Diaz
Page 95
[a] NORAD ENTRANCE, © NORAD, [b] SITE SECURITY PLAN, Stephen Valentine, Architect
Page 96
[a] TCV COMPONENT, Stephen Valentine, Architect, [b] COMPOSITE NANOSTRUCTURE, Ghim Wei Ho, and Professor Mark Welland, [c] MODEL: CRYO STORAGE LEVEL, Stephen Valentine, Architect
Page 97
[a] MODEL: COLD ROOM AIRLOCK, Stephen Valentine, Architect, [b] PARIS POMPIDOU CENTER, Stephen Valentine, Architect
Page 98
[a] WINDMILLS, Jamie Kripke/Getty Images, [b] ZEN ROCK GARDEN, Stephen Valentine, Architect, [c] WRIGHT'S TALIESIN WEST, Ezra Stoller © Esto
Page 99
[a] TIMESHIP POWER FACTORY, Stephen Valentine, Architect, [b] TURBINE ENGINE BLADES, © Corbis

CHAPTER SIX: SITE SELECTION
Page 101
TUSCAN VALLEY, Sime/G.Simeone
Page 102
EARLY SITE SKETCH, Stephen Valentine, Architect
Page 104
[a] TUSCAN LANDSCAPE 1, Age Fotostock, [b] TUSCAN LANDSCAPE-2, Owen Franken/Getty Images, [c] TIMESHIP SITE SETTING, Stephen Valentine, Architect
Page 105
[a] INFLUENCERS MAP, Stephen Valentine, Architect, [b] TUSCAN LANDSCAPE-3, S. Pitamitz/Getty Images
Page 106
[a] EARTHQUAKE BUILDING DAMAGE, © Corbis, [b] SNOWSTORM, Rick Werne, [c] HURRICANE: GULF OF MEXICO, NOAA
Page 107
[a] NATURAL DISASTERS MAP, Stephen Valentine, Architect, [b] TORNADO FUNNEL, Jim Reed/PR Inc
Page 108
[a] GLOBAL WARMING, Stephen Valentine, Architect, [b] WW-II HIROSHIMA BOMBING, © Associated Press, [c] MAN-MADE DISASTERS LOCAL MAP, Stephen Valentine, Architect
Page 109
[a] MAN-MADE DISASTERS MAP, Stephen Valentine, Architect, [b] MURRAH BUILDING, US Corp of Engineering
Page 110
[a] TIMESHIP SITE CAD-1, [b] TIMESHIP SITE CAD-2, [c] TIMESHIP SITE CAD-3, Stephen Valentine, Architect
Page 111
TIMESHIP SITE MODEL: PLAN VIEW, Stephen Valentine, Architect
Page 118
PHOTO: STEPHEN VALENTINE, Dudley Reed

SCIENTISTS IN SUPPORT OF CRYONICS

Cryonics is a legitimate science-based endeavor that seeks to preserve human beings, especially the human brain, by the best technology available. Future technologies for resuscitation can be envisioned that involve molecular repair by nanomedicine, highly advanced computation, detailed control of cell growth, and tissue regeneration. With a view toward these developments, there is a credible possibility that cryonics performed under the best conditions achievable today can preserve sufficient neurological information to permit eventual restoration of a person to full health.

The following individuals encompass all disciplines relevant to cryonics, including biology, cryobiology, neuroscience, physical science, nanotechnology and computing, ethics, and theology. [Note: Institution in parentheses indicates source of degree.]

GREGORY BENFORD, PH.D.
(Physics, UC San Diego) Professor of Physics, University of California, Irvine, CA

ALEXANDER BOLONKIN, PH.D.
(Leningrad Politechnic University) Professor, Moscow Aviation Institute; Senior Research Associate, NASA Dryden Flight Research Center; Lecturer, New Jersey Institute of Technology, Newark, NJ

NICK BOSTROM, PH.D.
Research Fellow, University of Oxford, United Kingdom

KEVIN Q. BROWN, PH.D.
(Computer Science, Carnegie-Mellon) Member of Technical Staff, Lucent Bell Laboratories (retired), Stanhope, NJ

MANFRED CLYNES, PH.D.
Lombardi Cancer Center; Professor, Department of Oncology and Department of Physiology and Biophysics, Georgetown University, Washington, DC

DANIEL CREVIER, PH.D.
(MIT) President, Ophthalmos Systems Inc., Longueuil, Qc, Canada; Professor of Electrical Engineering (retired), McGill University & École de Technologie Supérieure, Montreal, Canada

ANTONEI B. CSOKA, PH.D.
Assistant Professor of Obstetrics, Gynecology, and Reproductive Sciences, University of Pittsburgh School of Medicine Pittsburgh Development Center, Magee-Womens Research Institute, Pittsburgh, PA

AUBREY D.N.J. DE GREY, PH.D.
Research Associate, University of Cambridge, United Kingdom

WESLEY M. DU CHARME, PH.D.
(Experimental Psychology, University of Michigan) Author of *Becoming Immortal*, Rathdrum, Idaho

JOÃO PEDRO DE MAGALHÃES, PH.D.
University of Namur, Belgium

THOMAS DONALDSON, PH.D.
Editor, Periastron; Founder, Institute for Neural Cryobiology, Canberra, Australia

CHRISTOPHER J. DOUGHERTY, PH.D.
Chief Scientist, Suspended Animation Inc., Boca Raton, FL

K. ERIC DREXLER, PH.D.
Chairman, Foresight Institute, Palo Alto, CA

ROBERT A. FREITAS JR., J.D.
Author, Nanomedicine Vols. I & II; Research Fellow, Institute for Molecular Manufacturing, Palo Alto, CA

MARK GALECKI, PH.D.
(Mathematics, University of Tennessee), M.S. (Computer Science, Rutgers University), Senior System Software Engineer, SBS Technologies, NM

D. B. GHARE, PH.D.
Principle Research Scientist, Indian Institute of Science, Bangalore, India

BEN GOERTZEL, PH.D.
(Mathematics, Temple) Chief Scientific Officer, Biomind LLC, Columbia, MD

PETER GOURAS, M.D.
Professor of Ophthalmology, Columbia University, New York, NY

AMARA L. GRAPS, PH.D.
Researcher, Astrophysics, Adjunct Professor of Astronomy, Institute of Physics of the Interplanetary Space, American University of Rome, Italy

RAPHAEL HAFTKA, PH.D.
(UC San Diego) Distinguished Professor, University of Florida; Dept. of Mechanical & Aerospace Engineering, Gainesville, FL

DAVID A. HALL, M.D.
Dean of Education, World Health Medical School

J. STORRS HALL, PH.D.
Research Fellow, Institute for Molecular Manufacturing, Los Altos, CA; Fellow, Molecular Engineering Research Institute, Laporte, PA

ROBIN HANSON, PH.D.
(Social Science, Caltech) Assistant Professor of Economics, George Mason University, Fairfax, VA

STEVEN B. HARRIS, M.D.
President and Director of Research, Critical Care Research, Inc, Rancho Cucamonga, CA

MICHAEL D. HARTL, PH.D.
(Physics, Harvard & Caltech) Visitor in Theoretical Astrophysics, Caltech, Pasadena, CA

HENRY R. HIRSCH, PH.D.
(MIT) Professor Emeritus, University of Kentucky, College of Medicine, Lexington, KY

TAD HOGG, PH.D.
(Physics, Caltech and Stanford) Research staff, HP Labs, Palo Alto, CA

JAMES J. HUGHES, PH.D.
Public Policy Studies, Trinity College, Hartford, CT

JAMES R. HUGHES, M.D., PH.D.
ER Director, Meadows Regional Medical Center, Vidalia, GA; Director, Medical Research & Development, Hilton Head Longevity Center, Savannah, GA

RAVIN JAIN, M.D.
(Baylor) Diplomate, American Board of Psychiatry and Neurology, Los Angeles, CA

SUBHASH C. KAK, PH.D.
Department of Electrical & Computer Engineering, Louisiana State University, Baton Rouge, LA

BART KOSKO, PH.D.
Professor, Electrical Engineering Department, University of Southern California

JAMES B. LEWIS, PH.D.
(Chemistry, Harvard) Senior Research Investigator (retired), Bristol-Myers Squibb Pharmaceutical Research Institute, Seattle, WA

MARC S. LEWIS, PH.D.
(Clinical Psychology, Ph.D., University of Cincinnati) Associate Professor, Clinical Psychology, University of Texas at Austin

BRAD F. MELLON, STM, PH.D.
Chair of the Ethics Committee, Frederick Mennonite Community, Frederick, PA

RALPH C. MERKLE, PH.D.
Distinguished Professor of Computing, Georgia Tech College of Computing, Director, GTISC (GA Tech Information Security Center); VP, Technology Assessment, Foresight Institute, Palo Alto, CA

MARVIN MINSKY, PH.D.
(Mathematics, Harvard & Princeton) MIT Media Lab and MIT AI Lab; Toshiba Professor of Media Arts and Sciences; Professor of E.E. and C.S., MIT, MA

JOHN WARWICK MONTGOMERY, PH.D.
(D.Théol, Strasbourg), (LL.D., Cardiff) Professor Emeritus of Law and Humanities, University of Luton, England

MAX MORE, PH.D.
Chairman, Extropy Institute, Austin, TX

STEVE OMOHUNDRO, PH.D.
(Physics, University of California at Berkeley) Professor Computer Science, University of Illinois at Champaign/Urbana

MIKE O'NEAL, PH.D.
Associate Professor and Computer Science Program Chair, Louisiana Tech University, Ruston, LA

YURI PICHUGIN, PH.D.
Former Senior Researcher, Institute for Problems of Cryobiology and Cryomedicine, Kharkov, Ukraine

PETER H. PROCTOR, M.D., PH.D.
Independent Physician & Pharmacologist, Houston, Texas

MARTINE ROTHBLATT, PH.D., J.D., M.B.A.
Launched several satellite communications companies including Sirius and WorldSpace; Founder and Chief Executive Officer, United Therapeutics

KLAUS H. SAMES, M.D.
University Medical Center Hamburg-Eppendorf, Center of Experimental Medicine (CEM) Institute of Anatomy II: Experimental Morphology, Hamburg, Germany

ANDERS SANDBERG, PH.D.
(Computational Neuroscience, Royal Institute of Technology, Stockholm University, Sweden)

SERGEY V. SHELEG, M.D., PH.D.
Senior Research Scientist, Alcor Life Extension Foundation, Scottsdale, AZ

STANLEY SHOSTAK, PH.D.
Associate Professor of Biological Sciences, University of Pittsburgh, PA

RAFAL SMIGRODZKI, M.D., PH.D.
Chief Clinical Officer, Gencia Company, Charlottesville, VA

DAVID S. STODOLSKY, PH.D.
(University of California, Irvine) Senior Scientist, Institute for Social Informatics, Copenhagen, Denmark

GREGORY STOCK, PH.D.
Director, Program on Medicine, Technology, and Society, UCLA School of Public Health, Los Angeles, CA

CHARLES TANDY, PH.D.
Associate Professor of Humanities and Director of the Center for Interdisciplinary Philosophic Studies, Fooyin University, Kaohsiung, Taiwan

PETER TOMA, PH.D.
President, Cosmolingua, Inc., Sioux Falls, South Dakota; Inventor and Founder of SYSTRAN; Director of International Relations, Alcor Life Extension Foundation, Scottsdale, AZ

MARK A. VOELKER, PH.D.
(Optical Sciences, U. Arizona) Director, Bioengineering, BioTime, Inc., Berkeley, CA

ROY L. WALFORD, M.D.
Professor Emeritus, Pathology, UCLA School of Medicine, Los Angeles, CA

MARK WALKER, PH.D.
Research Associate, Philosophy, Trinity College, University of Toronto, Canada

MICHAEL D. WEST, PH.D.
President, Chairman & Chief Executive Officer, Advanced Cell Technology, Inc., Worcester, MA

RONALD F. WHITE, PH.D.
Professor of Philosophy, College of Mount St. Joseph, Cincinnati, OH

JAMES WILSDON, PH.D.
(Oxford University) Head of Strategy for Demos, an independent think-tank, London, England

BRIAN WOWK, PH.D.
Senior Scientist, 21st Century Medicine, Inc., Fontana, CA

STEPHEN A. VALENTINE, RA

One of internationally acclaimed architect Stephen Valentine's most significant design talents is his ability to integrate a bold symbolic vision with optimal functionality. His latest work is Timeship, a scientific think-tank and the world's center for the cryopreservation of the DNA of endangered species, human DNA, organs for transplant, and thousands of human patients traveling to the future. In this project, Valentine uses sacred geometry to create an edifice that poetically incorporates symbols of birth and re-birth that recur throughout history. He believes that a building must manifest its time, its program, and the aspirations of its builders, and that those encountering the building should viscerally experience these manifestations. Valentine is that rare architect of vision who is able to weave together an understanding of the mission of the project, an awareness of the significance of emerging technologies, a passion for detailing, and a mastery of the practicalities of construction to achieve a building in the tradition of the modern masters. But Timeship goes further to become a new, timeless architectural paradigm for the 21st century.

For more than two decades, Valentine has contributed to the design of major commercial and institutional projects worldwide. He was concept architect for a proposed new Long Island Railroad Station in New York City, located near the historic Grand Central Terminal. At I. M. Pei and Partners, he served as a senior architect for partner James Ingo Freed's highly acclaimed United States Holocaust Memorial Museum in Washington, D.C., and New York's Jacob Javits Convention and Exhibition Center, which has the world's largest space-frame. He was a senior architect for the Hong Kong Convention and Exhibition Center, the landmark structure that was the site for the transfer of governmental authority from the United Kingdom to the People's Republic of China in 1997. During his tenure as president and design director for the New York office of the Japanese firm Mirai International, Valentine led the design team for Superparadise, a multi-billion-dollar environmental science exposition park and habitat located outside Tokyo. He also was a member of a select team of international architects commissioned to design a master plan for the future development of the sacred city of Hangzhou, China, with a population of three million.

A graduate of Pratt Institute and recipient of the AIA School Medal, Valentine taught at Pratt Institute for more than a decade and has lectured to university groups, professional organizations, and governmental associations throughout the world.

ARTHUR L. CAPLAN, PHD

Arthur L. Caplan is Professor of Bioethics, Chair of the Department of Medical Ethics, and Director of the Center for Bioethics at the University of Pennsylvania. His interests include transplantation research ethics, genetics, reproductive technologies, health policy, and general bioethics. He is the author or editor of over 20 books and 500 papers on philosophy, medicine, health policy, and the biological sciences, and has served on national and international committees, including as Chair of the Advisory Committee to the United Nations on Human Cloning. His latest book is *Smart Mice, Not So Smart People,* in which he examines genetics, lifestyle drugs, the consequences of reversing aging, retooling the human brain, and cloning, explaining their impact and clarifying complex ethical issues.

ROBERT A. FREITAS JR

A pioneer in nanomedicine, Robert A. Freitas Jr., J.D., is Senior Research Fellow at the Institute for Molecular Manufacturing in Palo Alto, CA, and the author of *Nanomedicine*, the first book-length technical discussion of medical nanotechnology. He wrote the first detailed technical design study of a medical nanorobot ever published in a refereed mainstream medical journal. Research interests include medical nanorobotics, molecular machine systems, diamond mechanosynthesis, molecular assemblers, nanofactories, and self-replication in factory systems. Freitas co-authored *Kinematic Self-Replicating Machines*, another first-of-its-kind technical treatise, and has published more than 150 papers, book chapters, and popular articles on a wide range of scientific, engineering, and legal topics.

MICHAEL D. WEST, PHD

One of the founders of the fields of regenerative medicine and stem cell research, Dr. West has extensive academic and business experience in age-related degenerative diseases and telomerase molecular biology. Dr. West founded Geron Corporation, and from 1990 to 1998 served as a Director and Senior Executive Officer, where he initiated and managed research programs in telomerase diagnostics, inhibition, and therapy, as well as in human embryonic stem cells. He is now Chief Scientific Officer of Advanced Cell Technology where he leads its efforts in human stem cell research, and is also Adjunct Professor of Bioengineering at the University of California, Berkeley. His 2003 book, *The Immortal Cell*, tells the story of his philosophical, business, and scientific struggles with the problem of human aging.

JOHN LOBELL

John Lobell is a professor of architecture at Pratt Institute, where he addresses the intersection of architecture, history, culture, and technology. Lobell's interests range widely and include, besides architecture, mythology, information theory, and the impact of quantum and digital technologies on consciousness. He is currently working on the application of generative genomics, the process whereby simple rules can generate complex forms, to the fabrication of architecture. Lobell is the author of numerous articles and papers and several books, including *The Little Green Book,* a handbook for sustainability; *Joseph Campbell: The Man and His Ideas*; and *Between Silence and Light: Spirit in the Architecture of Louis I. Kahn*. He is currently working on a book titled *Architecture and Structures of Consciousness*.

ACKNOWLEDGMENTS

I want to thank those who have contributed to the technical development of the Timeship building, including: Vincent DeSimone, DeSimone Consulting Engineers; Robert Ducibella, Ducibella, Venter & Santore Consulting Engineers; Valentine Lehr, Lehr Consultants International; Michael Iarocci, Cryogenic Designer; Brent M. Porter, Professor of Architecture, Pratt Institute; Weidlinger Associates; Brian Wowk, PhD, Senior Scientist, 21st Century Medicine; and Seth Joseph Weine.

I also want to thank all those who contributed to or indirectly inspired the realization of this book, including Dorothy Anodide, my second grade teacher, formerly of Prescott School; John Lobell, Professor of Architecture, Pratt Institute; Gregory M. Fahy, PhD, Vice President and Chief Scientific Officer and Brian G. Wowk, PhD, Senior Physicist, both of 21st Century Medicine; Steven B. Harris, MD, President and Director of Research at Critical Care Research, Xi Zhao-Wilson, PhD, CEO and President at Biomarker Pharmaceuticals; Michael D. West, PhD, Chairman of the Board, Chief Scientific Officer, and former CEO of Advanced Cell Technology, Robert A. Freitas Jr., Senior Research Fellow at the Institute for Molecular Manufacturing; Arthur L. Caplan, PhD, Professor of Bioethics and Director of the Center for Bioethics at the University of Pennsylvania; Brenda Peters, Jenna Martin, and Philip Smith, advisors to the book; Massimo Vignelli, Vignelli Associates, graphic design advisor; and special thanks to the publishing staff at The Images Publishing Group, Mulgrave, Australia, especially Paul Latham and Alessina Brooks.